People's Reactions to Technology

The Claremont Symposium on Applied Social Psychology

This series of volumes highlights important new developments on the leading edge of applied social psychology. Each volume concentrates on one area where social psychological knowledge is being applied to the resolution of social problems. Within that area, a distinguished group of authorities present chapters summarizing recent theoretical views and empirical findings, including the results of their own research and applied activities. An introductory chapter integrates this material, pointing out common themes and varied areas of practical applications. Thus each volume brings together trenchant new social psychological ideas, research results, and fruitful applications bearing on an area of current social interest. The volumes will be of value not only to practitioners and researchers, but also to students and lay people interested in this vital and expanding area of psychology.

Books in the Series

Interpersonal Processes, *Stuart Oskamp and Shirlynn Spacapan, Editors*
The Social Psychology of Health, *Shirlynn Spacapan and Stuart Oskamp, Editors*
The Social Psychology of Aging, *Shirlynn Spacapan and Stuart Oskamp, Editors*
People's Reactions to Technology, *Stuart Oskamp and Shirlynn Spacapan, Editors*

People's Reactions to Technology

In Factories, Offices, and Aerospace

Stuart Oskamp
Shirlynn Spacapan
Editors

The Claremont Symposium on Applied Social Psychology

SAGE PUBLICATIONS
The International Professional Publishers
Newbury Park London New Delhi

For information address:

SAGE Publications, Inc.
2111 West Hillcrest Drive
Newbury Park, California 91320

SAGE Publications Ltd.
28 Banner Street
London EC1Y 8QE
England

SAGE Publications India Pvt. Ltd.
M-32 Market
Greater Kailash I
New Delhi 110 048 India

Printed in the United States of America

Library of Congress Cataloging-in-Publication Data

People's reactions to technology in factories, offices, and aerospace / edited by Stuart Oskamp and Shirlynn Spacapan.
p. cm. — (The Claremont Symposium on Applied Social Psychology ; v. 4)
Includes bibliographical references.
ISBN 0-8039-3852-7. — ISBN 0-8039-3853-5 (pbk.)
1. Machinery in industry—Congresses. 2. Automation—Social aspects—Congresses. 3. Business—Data processing—Social aspects—Congresses. 4. Aerospace industries—Automation—Social aspects. I. Oskamp, Stuart. II. Spacapan, Shirlynn. III. Series: Claremont Symposium on Applied Social Psychology (Series) ; v. 4.
HD6331.P43 1990
338.4'5—DC20 90-32540
CIP

FIRST PRINTING, 1990

Sage Production Editor: Kimberley A. Clark

Contents

Preface

This volume brings together chapters on one of the most pervasive issues of modern life—how people respond to the many kinds of technology that surround them and so completely affect their lives. This is a topic that applied social psychologists have studied from many aspects and with many different methods. Three types of technology are discussed in this volume: computerization of offices, use of robots and other computer-controlled machines in factories, and advanced technology in aviation and space exploration applications.

The chapters herein are expanded versions of presentations given at the sixth annual Claremont Symposium on Applied Social Psychology, held on February 4, 1989. This annual conference brings together outstanding psychologists from various sections of the country to join in the discussion and analysis of important topics and issues in the field of applied social psychology. The symposium has proven fascinating and informative to audiences of professional psychologists, students, and faculty members; and the chapters in this volume will be equally valuable and stimulating to readers, for they describe research at the cutting edge of this rapidly changing field.

We are grateful for financial support for the conferences from all six Claremont Colleges (Claremont Graduate School, Claremont McKenna College, Harvey Mudd College, Pitzer College, Pomona College, and Scripps College). We are also grateful to Dr. Rudi Volti for his thoughtful comments in moderating a panel discussion among the participants at the conference. Our personal thanks go to Shawn Okuda for her help with indexing and to Catherine Cameron and Mike Nichol for their support.

To provide integration of the content from the various presentations, we have included an introductory chapter that highlights key issues in people's responses to technology, summarizes some of the basic research findings, and draws together common themes that ran through several of the authors' presentations.

The expanded versions of the papers presented at the conference make varied and valuable contributions to the social psychology of technology. It is our hope that this volume will be interesting and useful to researchers and practitioners in the areas of both social psychology and technology, as well as to scholars who want to know more about applied aspects of implementing technology in various settings.

Stuart Oskamp
Shirlynn Spacapan
Claremont, California

1

People's Reactions to Technology

SHIRLYNN SPACAPAN
STUART OSKAMP

Technology is probably the most dominant influence on life in the modern world. Of course, technology isn't entirely new. Our ancestors had to adapt to the development of horse-drawn vehicles, sailing ships, steam engines, railroads, telephones, and automobiles. It is the *pace* of change in modern technology and its *pervasiveness* in all areas of our lives that are entirely new phenomena.

People have very different reactions to both the pace and pervasiveness of new technology in their lives, and that is the topic of this volume. In this chapter, we first sketch some of the ways in which new technology has affected our lives and then give a brief preview of each of the following chapters. Finally, we discuss several key themes that recur frequently in this volume and that reflect the most important principles and aspects of research on the human-technology interface.

Pace of Technological Change

Three areas of technology that have evidenced rapid change in recent years are those of computers, robots, and the aerospace industry. A few costs and benefits of new developments in each of these fields are discussed briefly below, and in greater detail in the chapters to follow.

Computers

Probably uppermost in many people's minds when they think of technology are computers. And it is no wonder, for recent data suggest that the rate of diffusion of personal computers has steadily increased in the last five years (McPadden, 1988). However, there are markedly conflicting reports of the value of computers in many areas of modern life. A report from a recent psychology convention proclaimed the great virtues of computers in college teaching, illustrating this conclusion with data from classes in experimental psychology (Bales, 1988a). The advantages reported included the following: Teachers can present more material in the same class time, study materials are available freely and easily at any hour of the day, students can progress at their own pace, tests can be conveniently given and scored by computer, and students learn more and get better grades.

On the other side of the ledger, another research report described many problems in the use of microcomputers in high school classes (Schofield & Verban, 1988). Among the major barriers to computer use, the report listed general inertia, minimal understanding of computers on the part of most teachers, the overload on the few computer-knowledgeable teachers, and a feeling of threat to their self-esteem that learning about computers poses for many teachers.

Computers also have an entertainment function. Just like VCRs, sports cars, cellular car phones, or stereo headsets that will play "music wherever you go," computers are "toys for adults" and status symbols that we use to impress ourselves and others (Adams, 1988).

Robots

In recent years, robots have taken over many factory jobs that involve repetitive and/or precise movements. On the other hand, a recent research paper reported that where robots have been introduced in factories, some are sitting unused since their implementation was not always a success (Fleischer, 1988). There are a variety of reasons for this difficulty in getting technology to be used effectively, and this volume discusses many aspects of both the problems and the promise of factory automation.

Outside of factories, many of us may soon become more familiar with robots. A brand-new breed of "service robot" is already in use in an increasing number of hospitals and office buildings, delivering meals to nursing units, distributing office mail, and making security patrols. Before the end of the 1990s, it is predicted that many homes will utilize similar robots—"kindly androids that will dust, scrub, vacuum and, yes, even do windows" (Koenenn, 1988).

Aerospace

In the aerospace area, extensive investigations have revealed that airline crashes often are due to problems relating to the interaction of humans with the complex aircraft technology. Also, at high speeds and under multiple stresses, problems in communication or interaction among the cockpit crew or with air controllers became especially dangerous. A recent article (entitled "Wild Blue Blunders") described some of the research findings in this area and concluded that "most accidents in the air are due to poor management in the cockpit" (Stark, 1988, p. 30).

Technological Disasters

Other catastrophe stories emphasize the same theme that human reactions to technology can be fraught with perils. In many catastrophes, there was even technology available to prevent a technological disaster, as in the Chernobyl nuclear meltdown. In that instance, the anticipated human reaction of fear kept plant officials from having employees practice and learn the emergency procedures that could have averted the disaster (see Kanki & Foushee, this volume, Chapter 6). In other cases, it is the sheer amount of technical information and related responsibilities that overwhelm human decision-makers and set the stage for impending crises. We all remember the tragic decision errors that led to the space shuttle *Challenger* explosion (including the decision to launch when the weather was too cold). More recently, the shooting down of an Iranian airliner by the U.S. missile cruiser *Vincennes* was attributed to the difficulties of dealing with masses of technological information in fleeting moments of time while under many other pressures. Four psychologists testified at congressional hearings that "findings from psychological research could have predicted or helped avoid some of the errors that led to the tragedy" (Bales, 1988b, p. 10).

This volume won't solve all these problems and questions regarding human interactions with technology. It does, however, discuss many aspects of these three major types of technology (computers, robots, and aerospace) and presents many principles derived from research that are applicable to people's dealings with technology.

Pervasiveness of Technology

An indication of the degree to which a given phenomenon pervades (or sometimes, invades) our lives can be derived from the ways in which the term crops up in our everyday language. For example, several writers have pointed out how the jargon formerly used only by computer hackers is now commonly found in reference to the self—we speak of being in a "work mode" when engaged in a consuming task; an interpersonal conflict with a colleague may now be called an inability to "interface"; when mentally exhausted, we apologize for a "slow processing speed"; we call our reflexive responses part of the "hardwiring" of our "system"; attending a business luncheon is part of our efforts to "network"; and, when left out of such luncheons, we feel we need to get back "in the loop" (see Turkle, 1984). References to technology in general have also crept into our language, as illustrated by the following four examples.

The "Technological Fix"

As many writers have pointed out, Americans are considerably more favorably disposed toward new technology than people in most other countries, including Japan (Volti, 1988). In our culture, it appears that most people welcome the promise of new technology as an asset in their lives. In commerce, fortunes are made by those who first produce talking dolls, advanced home video systems, or better lawn mowers or snowblowers. Occasionally, however, after people's experience accumulates, a technology may be rejected as being too ineffective or more of a torment than a help (e.g., computerized phone solicitation, video telephones, or electronic mail for executives). A key variable in determining acceptance of new technology seems to be whether users have control over the decision of whether and when to use it (see the section on personal control later in this chapter).

As a society, we seem to believe in the idea of a "technological fix." In the case of environmental crises, people seem to think that the solution "only requires the right technology" (Fisher, Bell, & Baum, 1984, p. 346). Rather than adopt personal conservation behaviors to cope with the problem of waning energy resources, we look to nuclear and solar technologies instead; rather than carpool or use public transportation to lessen air pollution, we install car-exhaust antipollution devices as a "technofix." In our personal lives, rather than adopt a healthy lifestyle, some of us fantasize that medical technology in the form of a new magic pill or treatment will allow us to "live as immoderately as we choose, and not contract either cancer or heart disease" as well as not to gain weight (Naisbitt, 1982, p. 53); we pray for a technical cure for AIDS, rather than practice safe sex and abandon a life of intravenous drug use. We even look to new technology to fix the problems created by previous technological innovations: We finally have a technological fix for the telephone—considered by some an "instrument of torment"—in the form of the answering machine. But that "solution" is in turn generating new sources of confusion and avoidance of interaction, and now we await a technofix for answering machines (Volti, 1989). Thus new technologies are generally welcomed in our society, and, at least initially, usually receive a "good press."

Technophobia

Some people do, however, fear, hate, loathe, or agonize about technology. One example of this is seen in our earlier description of the resistance by high school teachers to the use of microcomputers in their classrooms. Another example is reported by Nieva, Newman, and Bottlik (this volume, Chapter 3), who found that factory workers with more complex jobs were more accepting of additional, new technology than were employees with simpler jobs. A suggested explanation for this is that the latter group probably expected bad things (e.g., job loss) to happen as a result of the new technology.

A new term has been coined for these kinds of reactions – *technophobia* – and it is often bandied about in reference to any tendency to avoid advanced technology. An individual may, however, have several reasons for not using the new technology. Managers may avoid using their office computer because they are simply too busy to learn how, they believe it is somebody else's job to use it, or they find it to be a demeaning, clerical type of task. In other cases, a sophisticated piece of equipment may seem unnecessary when it really doesn't improve performance above and beyond the familiar "low tech" alternative. An intriguing illustration of this is the preference by astronauts for a roll of duct tape instead of a complicated vacuum-suction device to hold down small objects in a zero-g (gravity) work environment (see Wichman, this volume, Chapter 7). One could hardly accuse astronauts of being technophobic!

Whatever the underlying motivation for a reluctance to use new technologies, electronics manufacturers spend a great deal of effort trying to make their products attractive enough to surmount the obstacle of technophobia. An ad from AT&T declares that "even the most advanced technology is of little use if people are afraid of using it," so the company employs many psychologists who "observe people to test their reactions to new products. To find out what they're comfortable with. And what they're not" (*Time*, 1988, p. 51).

Technostress

Despite the efforts of advertisers and others concerned with making new technology "user friendly," there are many concerns about the health effects of the technology inherent in such areas as nuclear plants, pesticides, and supersonic transport. These real or anticipated effects, which cause concern, worry, and anxiety in some people, have been called *technostress* (Brod, 1984). Even one's own personal computer may be a source of technostress beyond just the usual frustrations people experience when "interacting" with this machine. For example, an apparent health epidemic of computer-related stress and actual physical pain, known as *repetitive strain injury* (RSI), is rampant among Australian clerical workers; it is so extensive that the resulting claims are threatening to overwhelm the nation's workers' compensation system. The fact that this complaint is rare among computer workers in many other industrialized nations illustrates how the social context of work and technology influences people's reactions to them (Kiesler & Finholt, 1988).

"High Tech – High Touch"

The idea that the social context of work and technology may in some way compensate for the costs we incur in embracing new technologies was suggested in the early 1980s (Naisbitt, 1982). In outlining the directions that our sophisticated technology would take us, futurist Naisbitt predicted that "high tech – high touch" would be one of the ten key transformations of our lives. In analyzing past developments, he noted that in some cases high tech has facilitated high touch (e.g., jet airplanes led to more face-to-face business meetings; the Pill led to a revolution in sexual lifestyles) while in other cases the two have appeared almost simultaneously (e.g., high touch quality circles moved into U.S. factories along with high tech robots; the hospice movement developed alongside the advent of life-sustaining medical equipment). In still other cases, high touch represented a backlash against high tech (e.g., ultra-complex heart transplants and advanced hospital equipment have led to renewed interest in the concepts of a family doctor, home birthing, and natural medicine; word processing has led to more handwritten notes; and high tech firms have turned to "management by wandering around" – Peters & Waterman, 1982). In more extreme cases, there has been outright rejection of "high tech" (e.g., laws prohibiting high tech polygraph tests in some states because they are "dehumanizing"; refusals to accept a corporation's automatic payroll deposit of one's wages and consequent lawsuits). High tech – high touch, claims Naisbitt, is a trend in our society that represents our efforts to seek the necessary balance between physical technology, the social context, and the spiritual aspect of our lives. This trend forms a backdrop against which all of the following chapters – indeed, all of the social psychology of technology – may be viewed.

Overview of Following Chapters

The authors of the chapters in this volume are all active researchers and experts in their particular aspect of the human-technology interface, and many of them have published recent books on human reactions to new technology in factories, offices, and aerospace. In this volume, they focus on the principles that can be derived from such research findings, and they discuss how these principles can be applied to many different technological settings.

Factories

Majchrzak and Davis provide a broad overview of flexible factory automation in Chapter 2. While they focus on the human side of this topic, their chapter is also noteworthy for its introduction to some of the technical aspects that may be unfamiliar territory to social psychologists. They review the four "classic" human issues involved—human resource practices, organizational structure and culture, employment and displacement, and implementation processes—and dispel any illusions readers may have as to the eventual reality of a paperless, workerless factory. As discussed in a later section of this chapter, these authors propose that a sociotechnical systems (STS) perspective may be best for integrating human and technological aspects of industrial changes, and they suggest that STS researchers should explore the identification of sociotechnical matches and mismatches, and how to facilitate matches in organizations. After describing several specific studies in this tradition, Chapter 2 concludes with advice to social psychologists who may be interested in adopting an STS approach.

In Chapter 3, the conversion of a manufacturing facility to a computerized "factory of the future" was the setting for a 3-year longitudinal study by Nieva, Newman, and Bottlik. The conversion was accompanied by major shifts in organizational culture, and both sets of changes were prompted by corporate desires to regain a competitive edge in the international marketplace. The authors studied changes in the number of jobs, the structure of tasks, and job quality. Data were obtained from lengthy interviews with key organizational personnel, archival records and two waves of surveys of nonsupervisory employees. Chapter 3 gives an overview of findings regarding job changes as a result of new technology and also provides an excellent follow-up to several points made in Chapter 2. For example, the authors found that fears about massive layoffs and a "workerless factory" were not fulfilled, and sizes of individual work groups shifted in both directions, with some becoming larger and others smaller than at the outset of this 3-year study.

Offices

In Chapter 4, Gutek and Winter report on a study of the effects of office computerization on the job satisfaction of employees. Because earlier research had revealed that computer use had no consistent, direct effects on job satisfaction, the authors focused on two possible mediators of the effects: "symbiotic interaction" (feelings that one works well with the computer as one's partner) and perceived control over the computer. They also hypothesized that several aspects of computer use, such as the individual's computer tools, skills, and experience, would be associated with job satisfaction. Drawn from 49 Southern California organizations, their sample of 89 computer-using work groups ranged from managerial to clerical units. Their descriptive results show that the extent of computerization in many office settings is only partial and that computerized surveillance by supervisors is only reported in a minority of work groups. Though symbiotic interaction, perceived control, and experience with computers were each related to job satisfaction, routineness of work was by far the strongest predictor of low job satisfaction. The authors conclude by discussing implications of these findings for integrating computers in office settings.

The research of authors Kraut, Fish, Root, and Chalfonte, presented in Chapter 5, aims to improve the communication technologies that work groups have available to them, particularly those groups that are geographically separate. The authors focus on the area of informal communication because it is an important, and often overlooked, mechanism for achieving both the task goals and social goals of groups. In one study of informal communication in an R&D (research and development) organization, they found that the vast majority (over 85%) of interactions in their sample were informal, and that such conversations were relatively unplanned, occurred between those who ordinarily were in close proximity, and were shorter than formal communications. The authors then undertook four additional studies that explored the perceived value of informal communication for participants and the effect of informal communication on person perception, ease of collaboration, and the success of collaborations. The results of these studies are reported in this chapter. A closing section describes two new telecommunication systems, VideoWindow and Cruiser, that were designed to create an artificial proximity for geographically separated users in order to facilitate informal communication.

Aerospace

In Chapter 6, Kanki and Foushee summarize research on crew factors in the aerospace workplace, and report on two of their projects. One study explored group dynamics aspects by examining the effects of a captain's personality on his crew's performance in a highly sophisticated flight simulator during five flight segments; it found that more effective crews were those led by captains who were both task- and person-oriented. The second project is an ongoing systematic field study of life and work in an undersea habitat, where crews live in isolation for weeks at a time. The findings will be applied in the design of new technologies for the aviation industry and for spaceflight.

In our closing chapter, Wichman reports the results of two other recent studies involving aerospace issues, the first of which investigated human cognitive and emotional responses to problems in computerized voice control of communication equipment. The findings will be useful to the U.S. space program in designing speech-recognition equipment control systems to free an astronaut's hands from routine, time-consuming tasks. In the second study, the videotaped movements of astronauts on two earth-orbital *Spacelab* flights were coded in an effort to understand inefficient aspects of task behavior during weightlessness and to facilitate the design of appropriate body restraints for zero-g (gravity) work environments. The chapter suggests several areas in which social psychologists and other behavioral scientists can contribute research and successful applications in the space program.

Themes Regarding Reactions to Technology

In this section, we briefly summarize key themes regarding the varied ways that people react to technology in their lives. These themes are illustrated in the chapters to follow, as well as in the applied social psychology research literature in general.

Technology Should Be Broadly Defined

A simplistic view of technology as some mere machine, toy, or magic box (as in our examples of technophobia and technofix) is not to be found in the following chapters. Moreover, many of the contributors to this volume suggest that physical technology is not to blame for many of the problems and crises we have mentioned in this chapter. For example, Majchrzak and Davis (Chapter 2) point out that 75% of the difficulties in the computerization of manufacturing have been attributed to human issues; similarly, in Chapter 6, Kanki and Foushee note that at least 65% of all aircraft accidents can be attributed to human errors and not to new technical equipment per se. In the aviation industry, many technical innovations have resulted in changed roles for crew members and in changes in their task and working environment, all of which, in turn, have been related to the well-publicized disasters on commercial flights. This observation leads to a more global view of "new technology" as encompassing the entire human-machine interface. Indeed, technology need not even include a machine aspect (see Chapter 5), since it includes "well-established procedures for scheduling meetings and writing reports . . . bumping into a colleague in the hall" (Kraut, Fish, Root, & Chalfonte) as part of communication technology. In addition, technology may refer to highly varied configurations of equipment and people, as evidenced in Gutek and Winter's view that computer systems should be considered as "sets of loosely bundled technologies" (Chapter 4). These broad views of technology go hand in hand with a rejection of the narrow view of technological determinism, the second major theme in this volume.

Technological Determinism

The viewpoint of technological determinism is that the introduction of any new technology has predictable consequences in shaping the structure and activities of organizations and individuals (e.g., Feigenbaum & McCorduck, 1983). This viewpoint sees the relationship between a technology (narrowly defined) and human behavior as unidirectional: Technology is a force that constrains people's behavior; therefore, the implementation of new technology will result in organizational change. The epitome of this viewpoint was captured in the theme of the 1933 Chicago World's Fair: "A century of progress: Science discovers, technology applies, man conforms" (Volti, 1989). Some technological determinists stress the increases in productivity and efficiency that will result from new workplace technology. Other more pessimistic determinists have emphasized the negative results of factory technology in increasing autocratic control by management, "de-skilling" and routinizing jobs, and reducing the autonomy of workers (e.g., Braverman, 1974).

The authors in this volume do not subscribe to either of these extreme views of technological determinism. Rather, they agree with Shaiken (1984) that the same technology can be used in ways that are effective and successful or ineffective and self-defeating. Technology does not automatically de-skill (lessen the skill value of) jobs; it can empower people, augment their skills, and create a more varied work environment that allows more input from workers. Neither does it necessarily improve the lot of workers; it can centralize office or shop floor operations and put extreme power in the hands of management.

As examples of broader perspectives, Majchrzak and Davis (Chapter 2) discuss the alternative viewpoints of "technological emergence" (where organizational change is attributed to the interplay among humans, technology, and the situation) and "interactive determinism" (a combination of emergence and determinism views). They point out that research guided by a deterministic perspective, or even by these two broader viewpoints, has not been useful to practitioners who must implement computers, robots, or other advanced machinery in the workplace. Chapter 3 by Nieva, Newman, and Bottlik provides an excellent demonstration that previous predictions about the impact of factory automation on employees were overly deterministic—the number of jobs, structure of tasks, and job quality all increased in some work groups and decreased in others, contrary to what technological determinists might predict. Similarly, Gutek and Winter (Chapter 4) observe that there was little technological determinism inherent in the office computer technologies they studied, and they conclude that the use of these systems typically "mirrors the existing work environment rather than alters it." In general, it seems that technology is best viewed as an intervening variable that combines with many other antecedent and intervening variables in influencing outcomes like productivity, efficiency, or worker satisfaction (see Gutek, Bikson, & Mankin, 1984).

An Open Systems Perspective

In contrast to a narrow deterministic approach, a broad open-systems perspective is evident in this volume. Systems theory, as explained and applied to organizations by Katz and Kahn (1978), is "basically concerned with problems of relationships, of structure, and of interdependence" (p. 22). The issue of interdependence is directly addressed in several chapters in this volume, especially in Majchrzak and Davis' discussion of the interrelationship of flexible factory automation and organizational structure/culture. These authors state that structural and cultural changes are required in order to implement new technology efficiently and that the new technology will have subsequent effects on the organization's structure and culture. This interdependence is also seen in Nieva, Newman, and Bottlik's discussion of the difficulty of attributing their results to technological (as opposed to organizational culture) changes that co-occurred at their research site.

The relationship between subsystems within an open system is an issue that Kanki and Foushee discuss at several different levels in Chapter 6. On one level, they note that the seemingly dramatic recent changes in a pilot's job are the result of many separate, often unrelated, changes in tasks. When accidents occur and hindsight is 20/20, it becomes apparent that the individual task changes were not considered in relation to one another, nor under any integrated system of planning for change. At a more macro-level, the authors note that a given aircraft's technology may not match that which is in use by a given air traffic controller, and both of these systems may be different from that system employed by other aircraft in the same airspace.

This macro-level view reminds us that an open systems perspective also includes the consideration of the environment surrounding the organization and the organization-environment relations (Katz & Kahn, 1978, pp. 30-31). Technology is not used in a vacuum and the social context within which it is introduced vitally affects its success. This point is evident throughout Chapter 5 as Kraut and his colleagues describe their efforts to design a physical communication technology that will duplicate the advantages of face-to-face informal social communication. Other examples may be found in Chapter 7, wherein Wichman describes why *Spacelab* was designed like earthbound rooms with a floor and ceiling, despite the fact that in the zero-g of space, one does not need a floor to walk on since one does not "walk." Thus, the contributors to this volume have adopted an open systems approach in many aspects of their work.

Sociotechnical Systems

A theme within the open systems perspective of this volume involves the sociotechnical systems (STS) approach, which emphasizes "joint optimization" of both the technical and social subsystems of an organization (see Emery & Trist, 1960). This principle was not recognized by Frederick Taylor (1947) and other early proponents of "scientific management," who strove to find the "one best way" to do a job in their efforts to maximize organizational efficiency. They treated the individual (and the social subsystem) as just another cog in the technical subsystem, whereas the STS approach posits that organizational performance is enhanced when there is a good match between the technical and social subsystems.

Majchrzak and Davis provide an overview of STS concepts in their chapter, and they conclude that this match is best when there is ultimate human control over key decisions, and the technology is used flexibly as an aid to the human operator. While none of the other chapters deal with STS directly, all of them emphasize the importance of both the human and technical aspects of their topic. This is evident, for example, in Wichman's discussion of the problems of the third *Skylab* crew, wherein he refers to the "system, which included the people and the apparatus," and also when Kanki and Foushee discuss the mismatch between workers, tasks, and the aviation workplace. In the work reported by Kraut and his colleagues, the social and task functions that a physical communication technology must support are considered equally important.

Group Processes

Social psychologists have long stressed two important functions involved in group activity: task or production functions, and social-emotional or group maintenance functions (e.g., Bales, 1958). Informal communication among face-to-face group members is a key process in group activity, providing guidance concerning tasks, feedback on progress, and sociable interchange that builds familiarity, mutual understanding and liking, and group cohesiveness. That is a major reason why telecommuting has not become more common in our computer age—most organizations still need to maintain informal face-to-face communication among their workers. Individuals working together (and also those playing together) need to coordinate their activities, and the right kind of technology can help in this process. Kraut and his colleagues discuss how communication technology might be able to contribute effectively to the coordination process when group members are dispersed and cannot interact face-to-face. Such technology, they posit, should be designed to duplicate the advantages of informal social communication; specifically, it needs to be easy to access and needs to allow visual "browsing" to find potential communication partners and to stimulate memory jogging about topics that should be discussed. When such technology is available, however, there will still remain an interesting question as to whether people will be willing to change their interaction habits in order to use it for communication at a distance, or whether they will still prefer to chat face-to-face with whomever they happen to meet (thus demonstrating a "high touch" reaction to such "high tech" equipment, as discussed earlier).

In aerospace activities, small-group communication and coordination are critically important because of the time pressures imposed by modern jet flight and the safety risk of multiple fatalities that can result from errors or misunderstanding. Kanki and Foushee cite examples of safety failures in commercial aviation that stem from such coordination problems. In long-term spaceflight, some group interaction issues can become even more important, such as adapting behavior to conditions of weightlessness, extensive training demands, privacy needs, ingroup-outgroup relationships between aircrew and ground controllers, and cross-cultural understanding between groups from different nations and cultures. Wichman's chapter presents several examples of how these problems have already affected U.S. and Soviet space missions.

Job Satisfaction

The job satisfaction of workers is a key dependent variable in organizational research and one that has been particularly stressed by social psychologists and the human relations movement in industrial management (e.g., Likert, 1967). As discussed above, technological advances can have either favorable or unfavorable impacts on workers and hence their job satisfaction. Among the favorable potential impacts of computer-controlled production in factories are expanded job classifications, broader skills training, less intensive supervision, and other related changes in organizational culture (Majchrzak, 1988). To avoid the potential unfavorable impacts and enhance job satisfaction, Shaiken (1984, pp. 272-273) has recommended that any new technology be used in a way that creates or maintains jobs, improves the conditions of work, develops the industrial base, and improves the environment. However, reactions on the part of workers have ranged from a warm welcome for computerized production equipment through a wary avoidance of the new technology to strikes and sabotage.

Many research findings demonstrate that advanced technology in and of itself has little direct effect on job satisfaction. In office computerization, for instance, Gutek and Winter report that the routineness of work is the strongest factor in predicting low job satisfaction. Despite wide differences in the objective qualities of the hardware and software involved in office computer systems, most office workers report their equipment as being relatively helpful in their jobs, and they experience a high level of perceived control over it. Thus, it is not the objective technology, but rather the workers' perceived interactions with it, that predict outcomes such as job satisfaction.

Personal Control

One of the reasons why people may resist the introduction of new technology is that sophisticated technologies can threaten their perceptions of control over their lives. To take an extreme example, research has shown that technological disasters threaten people's perceptions of personal control even more than natural disasters, despite the fact that natural disasters are the ones over which they really have no control (Fisher et al., 1984, pp. 92-96). Another extreme example, provided by Kanki and Foushee, is the dependence of astronauts on life-support systems, which gives astronauts very little environmental control over such dimensions as temperature, lighting, living space, noise, and variety. Wichman describes an even worse potential loss of control that would be experienced by space station crews who, if an emergency occurred, might have to wait as long as four weeks before a shuttle could be readied to bring them back to earth.

Closer to home, several chapters herein note that workers' feelings of control over physical technology are important to the success of that technology in the workplace. The perception of control that workers have over their computer was a major variable of interest in the research of Gutek and Winter. Similarly, Majchrzak and Davis report the results of a recent study that found diminished feelings of autonomy among research and development designers after the implementation of a computer-aided design system. In the STS approach espoused by Majchrzak and Davis, control over key decisions about the tasks (with the computer as a helpful tool) must remain in the hands of the human in order to enhance the chance of a successful sociotechnical match. In the aerospace realm, Kanki and Foushee note that ultimate human control over normally automated functions has been found necessary in order to lessen the risk of disaster when emergencies occur. In the research of Nieva and her colleagues, a key aspect of the plant's gains in competitiveness involved changes in organizational culture, changes that emphasized each employee's personal responsibility for product quality. Finally, future success of the experimental communication technologies described by Kraut and his colleagues will depend on the proper balancing of perceived control needs. This balance, an inevitable aspect of privacy concerns (see Altman, 1975), is one that pits human desires for control of other people's access to oneself against the easy access required by informal communication.

Benefits and Costs of Technology

Advanced technology can greatly increase factory productivity, as illustrated in the results reported by Nieva et al. Computerization of offices can similarly increase the efficiency of white-collar work (see Gutek & Winter), and advanced aviation technology has dramatically increased the speed, smoothness, and safety of aerospace operations (see Kanki & Foushee).

However, there are many costs for all these advances, the most obvious being the capital expense of the new equipment. It is usually known in advance and evaluated as less than the resulting benefits; otherwise, the new technology would not be adopted. The many other costs are less quantifiable, and some of them are often overlooked completely. First, there are costs of training workers to use the new equipment and procedures; these can usually be estimated at least roughly based on past experience, but the related costs of lowered productivity during the early stages of using the new technology are harder to estimate accurately. A familiar example of these problems is the computerized accounting systems or management information systems that fail to work properly for months after their introduction.

When new equipment requires workers with different or greater skills, another possible cost is employee payroll increases rather than the hoped-for decreases. When workers are displaced by new technology, there are substantial costs for retraining them for other jobs or relocating them geographically. If employers themselves do not cover these costs, they fall on society in the form of welfare, unemployment, or retraining payments.

Finally, there are important but intangible costs of implementing new technology that are rarely ever considered—namely the social and psychological costs in terms of lower employee job satisfaction, motivation, absenteeism, turnover, complaints, and grievances. As mentioned above, the most extreme form of these costs can involve strikes and sabotage, but even less extreme reactions of disaffection can create heavy and unexpected financial burdens. Other aspects of psychological costs stemming from office communication technology include issues of personal privacy and control over the technology versus free access to coworkers when needed (see Kraut et al.). Many of the costs of technology are touched on by various chapters in this volume.

Universal Problems and Research Methods in Unique Environments

A variety of unique environments form the settings of the studies reported in this volume. The investigations reported by Kraut and his colleagues all involve research and development workers in such settings as a state university or an industrial research laboratory. Videotapes of astronauts in earth orbit provided Wichman with the data for one of the studies reported in his chapter, and studies of aviation flight crews and undersea habitats are reported by Kanki and Foushee.

However, the contributors to this volume point out that problems at the human-machine interface are not unique to their specific environments. Kanki and Foushee discuss five problem areas that are common to any workplace experiencing the rapid influx of technical changes: individual-level factors of personality differences and physiological demands, interpersonal processes and group dynamics, redefined tasks and environmental demands, organization and management problems, and training issues. Similarly, the issues that Majchrzak and Davis list as classic ones in the study of human aspects of implementing flexible factory automation are familiar areas of concern in studying the implementation of *any* organizational change. In the area of factory automation, a crucial question that must be considered is how best to define and measure organizational performance (see Majchrzak & Davis). Related issues concerning the many different benefits and costs of technology are raised by Nieva et al.

The integrated research approach espoused by Kanki and Foushee is also one that is applicable universally—not just in their unique environment of the aerospace workplace—because different research paradigms and methods have complementary strengths and weaknesses. In the area of office technology, the chapter by Kraut et al. illustrates how a wide variety of research approaches can be used to build a coordinated picture of the virtues and problems of innovative technologies. Gutek and Winter's chapter illustrates the importance of including intervening psychological and experimental characteristics as well as task variables in studying the effects of technology on outcome variables like job satisfaction. Thus another common thread that runs through the following chapters is the universality of the problem areas and research methods employed, across a surprisingly diverse set of environments.

The following chapters provide fascinating glimpses of the wide array of methods and findings that applied social psychologists have contributed to research on important questions of the use of technology and people's reactions to it. Numerous opportunities for additional research are suggested throughout this volume, for applied social psychology has much to contribute to understanding the uses, effects, and costs of technology.

References

Adams, J. L. (1988, Summer). Big kids' toys: Technology as entertainment. *The Stanford Magazine*, pp. 40-44.

Altman, I. (1975). *The environment and social behavior*. Monterey, CA: Brooks/Cole.

Bales, J. (1988a). Computers in the classroom. *APA Monitor*, *19*(6), 34.

Bales, J. (1988b). Vincennes: Findings could have helped avert tragedy, scientists tell Hill panel. *APA Monitor*, *19*(12), 10-11.

Bales, R. F. (1958). Task roles and social roles in problem-solving groups. In E. Maccoby, T. M. Newcomb, & E. L. Hartley (Eds.), *Readings in social psychology* (3rd ed.). New York: Holt, Rinehart & Winston.

Braverman, H. (1974). *Labor and monopoly capital: The degradation of work in the twentieth century*. New York: Monthly Review Press.

Brod, C. (1984). *Technostress*. Reading, MA: Addison-Wesley.

Emery, F. E., & Trist, E. L. (1960). Socio-technical systems. In *Management science models and techniques* (Vol. 2). London: Pergamon.

Feigenbaum, E. A., & McCorduck, P. (1983). *The fifth generation: Artificial intelligence and Japan's computer challenge to the world*. Reading, MA: Addison-Wesley.

Fisher, J. D., Bell, P. A., & Baum, A. (1984). *Environmental psychology* (2nd ed.). New York: Holt, Rinehart & Winston.

Fleischer, M. (1988, August). *Work organization and advanced technology in manufacturing*. Paper presented at the meeting of the American Psychological Association, Atlanta.

Gutek, B. A., Bikson, T. K., & Mankin, D. (1984). Individual and organizational consequences of computer-based office information technology. In S. Oskamp (Ed.), *Applied social psychology annual 5: Applications in organizational settings* (pp. 231-254). Beverly Hills, CA: Sage.

Katz, D., & Kahn, R. L. (1978). *The social psychology of organizations* (2nd ed.). New York: John Wiley.

Kiesler, S., & Finholt, T. (1988). The mystery of RSI. *American Psychologist*, *43*, 1004-1015.

Koenenn, C. (1988, October 17). At home with robots. *Los Angeles Times*, View Section, p. 1.
Likert, R. (1967). *The human organization: Its management and value*. New York: McGraw-Hill.
Majchrzak, A. (1988). *The human side of factory automation*. San Francisco: Jossey-Bass.
McPadden, J. (1988). *The rate of diffusion of the personal computer in the office environment*. Unpublished doctoral dissertation, Claremont Graduate School, CA.
Naisbitt, J. (1982). *Megatrends: Ten new directions transforming our lives*. New York: Warner.
Peters, T. J., & Waterman, R. H. (1982). *In search of excellence*. New York: Harper & Row.
Schofield, J. W., & Verban, D. (1988). *Barriers and incentives to computer usage in teaching*. (Tech. Rep. No. 1). Pittsburgh: University of Pittsburgh, Learning Research and Development Center.
Shaiken, H. (1984). *Work transformed: Automation and labor in the computer age*. New York: Holt, Rinehart & Winston.
Stark, E. (1988). Wild blue blunders. *Psychology Today*, *22*(10), pp. 30-32.
Taylor, F. W. (1947). *Scientific management*. New York: Harper.
Time. (1988, May 14). To cure technophobia you need a good psychologist. p. 51.
Turkle, S. (1984). *The second self: Computers and the human spirit*. New York: Simon & Schuster.
Volti, R. (1988). *Society and technological change*. New York: St. Martin's Press.
Volti, R. (1989, February). *Problems and prospects in people's use of advanced technology*. Panel discussion presented at the Claremont Symposium on Applied Social Psychology, Claremont, CA.

PART I

FACTORIES

2

The Human Side of Flexible Factory Automation: Research and Management Practice

ANN MAJCHRZAK
DONALD D. DAVIS

It is nearly impossible to escape the discussion in the popular press of the global competitive battle in which American companies are locked. Firms from throughout the world vie for the same markets and customers. A key ingredient for their success at this effort requires considerable manufacturing flexibility, the flexibility that is made possible by a host of new managerial and manufacturing methods.

In this chapter, we first define flexible factory automation and the human concerns associated with its implementation, and then describe approaches used to study and manage these human concerns. We argue that current perspectives unnecessarily limit both research and management practice concerning human issues and flexible factory automa-

AUTHORS' NOTE: Preparation of this chapter was supported by Grant DMC-8713951 from the Division of Design, Manufacturing, and Computer-Integrated Engineering, National Science Foundation, and from a faculty innovation grant from the University of Southern California awarded to Ann Majchrzak.

tion. We explore the reasons for these limitations and recommend a preferred approach to research and practice—sociotechnical systems design. Finally, this chapter concludes with a discussion of how social psychologists might begin to contribute to this field.

Flexible Factory Automation

Flexible factory automation (FFA), also known as advanced manufacturing technology (AMT), refers to the use of reprogrammable software and hardware to manufacture products. The distinguishing feature of this new technology is its great flexibility. With this technology, it is easy to vary the type of part produced and the materials used to produce these parts, as well as the manner in which these parts are produced. It is also relatively simple to change the volume of orders processed and the size of batches produced. These flexibilities are achieved by implementing any number of different technologies. The least technological flexibility is gained by installation of simple, small stand-alone equipment capable of minor amounts of reprogramming with substantial effort. For example, a robot dedicated to programming one type of welding operation will not achieve substantial flexibility for the organization. At the other extreme are technologies that are so flexible as to produce just about anything within a designated amount of work space. For example, the computer-integrated manufacturing system at the IBM Lexington plant can manufacture typewriters, printer motors, frames, paper feeds, covers, keyboards, and electronic cards—just about any product, the company claims—within a 26 × 22 × 18-inch work space (Waterbury, 1986). The software driving the manufacturing equipment is simply reprogrammed in order to produce a different product.

Several technologies are included within the domain of flexible factory automation. Some of these technologies may be familiar, while others may not be so easy to recognize; all probably seem like an alphabet soup of acronyms. FFA generally consists of automated equipment and automated methods for controlling and managing this equipment. Computer-numerical-control machines (CNCs) are stand-alone, single machines that are software controlled. When joined together and controlled by a central computer, they form a network of distributed-numerical-control machines (DNCs). Both CNCs and DNCs are a comparatively primitive form of FFA that became available during the 1960s. Robots (programmable stand-alone devices) may be used for picking up, placing, or assembling parts. They may also be used for activities that are too hazardous or uncomfortable for humans, such as welding and painting. Automated storage and retrieval systems (ASRs) electronically control the delivery of raw materials and components to production equipment, as well as the removal and storage of completed parts. Storage and retrieval are often done with small, pilotless vehicles (AGVs, or automated guided vehicles) that move along designated paths, usually embedded in the factory floor. When multiple technologies such as robots and CNCs are used together, they are referred to as computer-automated manufacturing (CAM).

FFA also includes computerized functions that control, manage, and integrate manufacturing processes. Computer-aided design (CAD), a process of electronic drawing that allows easy redrawing, storage, and retrieval of drawings, as well as joint or iterative drawing by multiple designers, may be joined electronically with manufacturing equipment (the combination is called CAD/CAM). Electronically stored designs are sent in digital form to the shop floor, where they guide the movements of equipment during the manufacturing process. Multiple resource planning systems (MRPs) computerize inventory and production-scheduling functions. When computerized production equipment is integrated via electronic linkages, the firm moves toward computer-integrated manufacturing (CIM). When all manufacturing functions are electronically integrated, as shown in the wheel developed by the Society of Manufacturing Engineers in Figure 2.1, CIM has been achieved.

The technologies discussed above are representative of the areas in Figure 2.1. For example, AGVs and robots are examples of factory automation, CAD is an example of product/process, and MRP is an example of manufacturing planning and control. Each of these, in turn, derives from higher-level management activities, such as strategic planning and marketing. Most important, each of these is centrally controlled and integrated through the electronic exchange of information and the sharing of common data bases. This use of computerized information to integrate and control manufacturing efforts is the chief distinguishing feature of CIM. This integration is the key to manufacturing flexibility and all of its associated strategic advantages (see Jelinek & Goldhar, 1986). Due to the huge expense and complexity

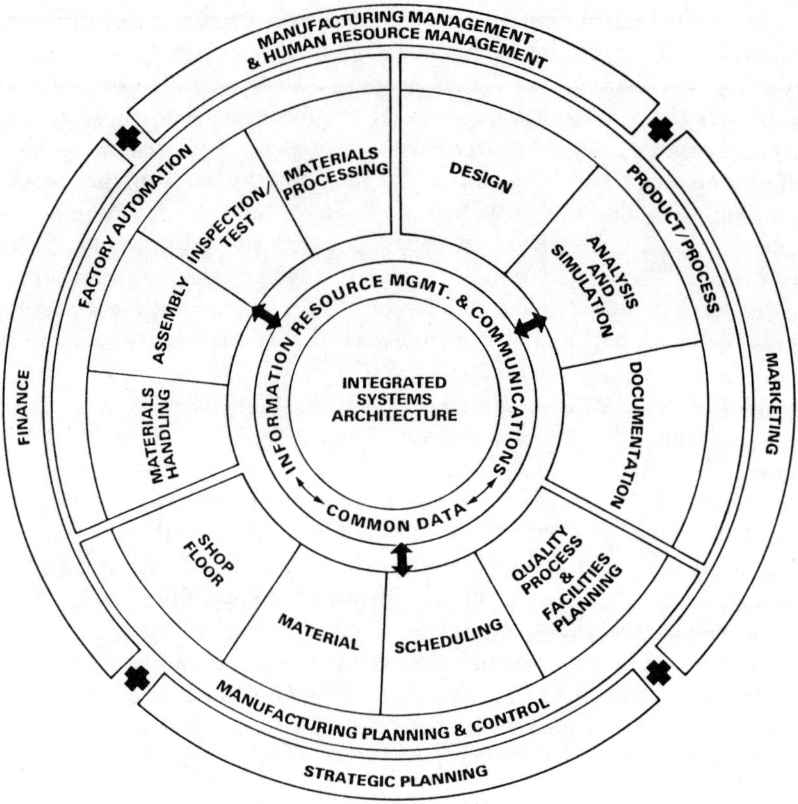

Figure 2.1. Integration of Manufacturing Functions

SOURCE: CIM "Wheel" developed by the Technical Council of the Society of Manufacturing Engineers' Computer and Automated Systems Association (CASA/SME) – copyright 1985 SME Second Edition, Revised November 5, 1985.

involved, as well as the characteristics of managers and their organizations, few plants have achieved CIM (Majchrzak, Nieva, & Newman, 1986). Unfortunately for U.S. interests, CIM is being used more often by foreign manufacturers, particularly – and unsurprisingly – in Japan.

Due to its reprogrammability, CIM is far more flexible than other forms of automation, such as traditional assembly lines. The capacity for using the same equipment to produce different parts with different sizes and features allows for immense manufacturing variability. The same production equipment can be used to produce goods in small batches of 50 or 100 or in large batches of thousands or millions. This flexibility offers a tremendous competitive advantage over traditional assembly lines. Furthermore, computer control allows almost perfect standardization, leading to great improvements in quality control. Production is more precise and consistent.

Traditional assembly lines, on the other hand, are economical only with production in very large batches, usually in the thousands or millions of parts, when economies of scale are reached. There is little flexibility. The entire assembly line must be retooled to produce a new part. This retooling typically costs millions of dollars and takes months to complete. Moreover, traditional assembly lines still require workers to do most of the physical labor. While the assembly line may deliver the part, workers must still bolt, cut, sand, paint, and assemble. Consistency of product quality is difficult to achieve since its dependence on workers creates fluctuations as the energy of the workers waxes and wanes.

FFA is radically different from traditional forms of automation; it is not simply an extension of the past. Its unique nature poses special demands on the organization and management of work, and failure to recognize that uniqueness limits the success of firms using this new technology. Foremost among these concerns is management of the people who must use these systems. Human issues are important since the success of these new manufacturing systems is limited by the abilities of those who manage and use them. The quality of the labor force in the firm (managers and workers) determines the upper limit of organizational potential and competitive advantage (Davis, 1986a).

Human Issues and Flexible Factory Automation

Labor savings historically have been the primary justification for the acquisition of automated equipment. It is believed that machines can replace workers, reducing the labor content and associated cost of manufactured goods. A common fantasy among managers in the U.S.

is the paperless, workerless factory that hums unstopped throughout the day and night, churning out products of high quality and low cost, without the problems resulting from the intrusion of people. This wish is unlikely to be achieved for some time (Ayres & Miller, 1983; Jaikumar, 1986; Parsons, 1988). For example, operators will still be needed to recognize complex patterns, react to unexpected events, generate creative alternatives, and solve unstructured problems – tasks that are impossible for FFA (and perhaps any machine technology) to achieve alone. People must design, implement, monitor, and maintain the system. People provide the limits of the technical system; it can work only as well as those who manage it.

Given the need for humans, it is quite important to discuss the human issues associated with FFA. Most who work in this field focus on the following topics, which are discussed briefly below:

- Human resource practices
- Organizational structure and culture
- Employment and displacement
- Implementation of new technologies

Human Resource Practices

Human resource practices include pay, training, job design, selection and promotion policies; union-labor relations; skills assessment and training; and assessment of job attitudes, satisfaction, and quality of work life. Many observers of FFA technology have warned that human resource practices used to manage conventional, nonflexible equipment are not appropriate for managing FFA (e.g., Majchrzak, 1988a; Manufacturing Studies Board, 1986; Office of Technology Assessment, 1984). For example, worker performance is traditionally evaluated and rewarded on an individual basis. With FFA, where production frequently requires a team effort, individually based evaluation and reward allocation become counterproductive at best and impossible at worst. Team-based performance, often across departments and shifts, must instead be measured and rewarded, requiring new methods of supervision and management.

Organizational Structure and Culture

Organizational structure includes the ways in which departments are differentiated, rules and procedures are formalized, decision-making authority is allocated, and people and departments are integrated vertically and horizontally. Organizational culture refers to the collective beliefs and attitudes of organizational members regarding the purpose, priorities, and unique identity of the organization. Observers of FFA technology have suggested that FFA will create a need to reallocate decision-making authority, restructure departments, increase formalization of some procedures, and change the way work activities are coordinated (Ettlie, 1988; Majchrzak, 1988a; Zuboff, 1988). These organizational and managerial changes will be required to manage the new technology effectively and efficiently. For example, workers in traditional factories, when confronted with a flawed part, are required to locate a supervisor with the appropriate authority to shut down the assembly line. During the interim, only a few inferior parts might be produced. With new forms of automation, however, computer integration amplifies errors. A mistake in one portion of the system sends waves of errors throughout the system. Within minutes, a program bug, for example, can cause damage in the hundreds of thousands of dollars. Workers must have the authority to stop production immediately when confronting such errors. The time needed to seek out proper authority will result in too much of a costly loss.

Some observers have gone so far as to suggest that CIM itself is a cultural change for the organization that adopts it; integration, coordination, quality, and responsiveness are given a much higher priority in a CIM factory than was previously necessary with conventional equipment (Adler, 1989; Roitman, Liker, & Roskies, 1988). For example, management communication and decision making must be at least as integrated as the production equipment itself. Information must be shared across departments, functions, levels, and shifts. Furthermore, production must be perceived in its entirety; managers must think systemically and not be unit-bound in their attention. The interconnectedness of everything increases mutual dependence. Traditional automation does not require this broad understanding among those who use it.

Employment and Displacement

Many have grave concerns about the restructuring of the work force in the United States today. Primary among these concerns is the potential for massive layoffs due to technological improvements. Hunt and Hunt (1983) estimate that each robot displaces about two jobs. This displacement is particularly likely in functions such as welding, painting, and loading/unloading. As many as 45% of all jobs in the United States could be affected by office or factory automation, with major consequences for individual workers (Office of Technology Assessment, 1984).

In contrast to this point of view, others argue that although many jobs that physically touch the product-in-process will be lost, new jobs off the shop floor will be created. Additional jobs in programming, engineering, and customer support will be needed that will often be best filled by those previously touching the product. The U.S. work force is sure to change, but fears of massive job loss in the manufacturing sector are grossly inflated.

Also of concern is the retraining of employees that must occur when tasks are displaced by the new equipment. Long-time employees will need to learn more than just new methods for operating the innovative equipment. Employees will experience a dramatically expanded work role. They will have to learn to visualize the entire production system, use statistical inference related to quality control procedures, work interdependently, and become adept in abstract thinking and problem solving (Davis, Roitman, Carlopio, & Fleischer, 1987).

Implementation of New Technologies

A final human issue of concern is the implementation of the new automation, particularly the factors that facilitate and impede the implementation of FFA. Several attempts have been made to describe the factors important in the implementation of FFA (Davis, 1986b; Dean, 1987; Ettlie, 1986; Mohr, 1987). Despite their differences, these discussions of implementation have in common a focus on people and their interaction.

Successful implementation of such new technologies as FFA requires an amalgam of important elements at multiple levels of analysis.

At the top of organizations, there must be an integration between the strategies that focus on technology, marketing, manufacturing, and human resources (Davis, 1986c). At middle and lower levels, managers must be prepared for changes in their work roles due to the new technology. For example, they must be willing to support team approaches to management and to loosen the reins of their power and authority. Throughout the organization, implementation can be facilitated by "synchronous innovation," the simultaneous and interdependent innovation and change in technology and management practices (Ettlie, 1988).

Study of Human Issues and Flexible Factory Automation

Technological Determinism

The approach used most commonly to guide the study of human issues in FFA has been the technological imperative, or technological deterministic model. This approach "views technology as an exogenous force which determines or strongly constrains the behavior of individuals and organizations" (Markus & Robey, 1988, p. 585). With such an approach, research tries to identify aspects of the human/organizational/social environment that are changed when FFA is installed. Majchrzak (1988a) has recently reviewed the studies conducted with this perspective. These studies often take the form of surveys that ask managers in manufacturing firms about the human resource changes they have made with the introduction of FFA, or they are case studies of the human resource changes made when an individual firm implements FFA.

An example of a study conducted within this approach is a set of interviews of workers concerning their reactions to the implementation of a robot in their workplace (Argote, Goodman, & Schkade, 1983). The researchers found that the robot increased interaction between operators and engineers and increased the mental workload of operators. Thus, in support of a technological determinist position, this study clearly demonstrated that the introduction of FFA has identifiable effects on humans.

Technological Emergence

In contrast to the technological deterministic approach is a view that assumes the uses and consequences of new technology emerge from complex social interactions involving people, roles, and objectives, and not from the technology itself. This approach has been characterized by Markus and Robey (1988) as the emergent perspective, and by Kling and Scacchi (1982) as a web model. In this approach, the technology is perceived as an ensemble of equipment, applications, and techniques that carry social meanings. Rather than attribute organizational change to the technology, this perspective emphasizes the dynamic interplay among workers, managers, technology, and organizational context. A study by Gasser (1986) that examined how engineer-users adapt to the idiosyncrasies of a computerized analytic program illustrates this perspective:

Gasser studied the integration of computing into routine work in ten public- and private-sector organizations. He described an episode that illustrates how people can cope with misaligned technical systems:

> Engineering analysts working in WESCO, a large firm which designed and constructed chemical plants, learned which analyses could go wrong in the package of complex analytical programs they used, and how to correct for untenable results. For example, they routinely input temperature coefficients for pipes carrying hot fluids as though the pipes were intended to operate cold, causing the analysis program to disregard certain heating stress calculations. Over years of experience with this program, the engineers had learned that entering the "correct" information would lead to erroneous results and the pipes would not work properly in the final design. They worked around the technical problems of the program by "running the hot pipes cold" when using it. (p. 206)

Interactive Determinism

A third approach to the study of human issues in FFA is a hybrid of the two views described above. This hybrid approach, which we call interactive determinism, presumes that there are some aspects of the organizational and human environment that change as a result of the technology, while there are other aspects that emerge from social interactions and cannot be predicted from knowledge of the technology alone. The important task within this perspective is to determine which organizational and human aspects respond in which manner. A study by Majchrzak (in press) of the implementation of a CAD system illustrates this approach. In the first set of analyses, she found that CAD seemed to create changes in research and development (R&D) designers' feelings of autonomy and teamwork. A second set of analyses (Majchrzak, Mosher, & John, 1988) found that one affected area was the manner in which R&D designers interacted with manufacturing personnel. The discussions of non-CAD users with manufacturing personnel were primarily routine and conducted in a fashion similar to the sequential nature of traditional design work. Discussions of CAD-users were more reiterative and unpredictable, with the perceived joint needs of manufacturing and R&D being shared. From these two analyses, it can be concluded that CAD has definable effects on perceptions of job autonomy and teamwork; however, some of these effects, such as conversations with manufacturing personnel, are clearly a function of how the organization uses the technology to foster engineering-manufacturing discussions.

A study by Yoon (1988) offers another example of the interactive determinist perspective. Yoon's study involved a survey of 185 Pennsylvania metalworking factories and found that while the installation of FFA was a significant factor in predicting increased decentralization of operating decisions in the factories, this decentralization did not translate into a reduced number of hierarchical levels in the organizations, as would have been expected. Managers in these firms maintained their positions despite having less work to do. Thus, the need to operate FFA effectively may force managers to give up some decision-making authority, but the decision of whether to give up their jobs and reduce the number of hierarchical levels is a function of nontechnological factors—in this case, the vested interests of those affected by the technological change.

These studies demonstrate the usefulness of an interactive determinist perspective. The complexity involved in studying these issues should also be apparent. As in the classical Gestalt portraits, figure and ground can shift, leading to different views of complex and ambiguous events.

Limitations of Current Approaches to Studying Human Issues

The perspectives currently used to examine human issues of FFA have guided a substantial number of research studies (for example, Majchrzak's 1988a review lists over 200 empirical studies of FFA). This research has in turn yielded a significant number of conclusions about the human issues associated with FFA. First, adopting integrated FFA creates more human changes in the organization than adopting stand-alone automated equipment. Second, FFA provides an opportunity for the adoption of broader job classifications, broader skills training, and group-oriented job designs. Third, alternative pay policies, patterns of supervision, and substantial retraining often are needed to meet the needs of the new job designs. Fourth, changes in organizational culture often must occur if the benefits of FFA are to be achieved. Finally, the implementation of FFA must emphasize user participation and preplanning (see Majchrzak, 1988a, for a more detailed discussion of these conclusions).

Despite the availability of the above rather firm research conclusions, these findings are seldom used by managers to guide their practice when implementing FFA. This disappointing state of affairs has been noted by the Office of Technology Assessment (1984), which concluded that "the main stumbling blocks in the near future for the implementation of [FFA] are not technical, but rather are barriers of cost, organization of the factory, availability of appropriate skills, and social effects of these technologies" (p. 94). Between 50% and 70% of the FFA installations implemented in manufacturing firms fail to achieve their intended benefits; a large portion of these failures are due to management's inability to handle the human issues effectively (Works, 1987). In a survey of manufacturing clients conducted in 1986 by the Yankee Consulting Group (Criswell, personal communication, 1988), 75% of the difficulties of implementing computers in manufacturing were attributable to organizational and people issues.

An important paradox is apparent. At the same time that FFA implementation typically fails due to mismanagement of human and organizational concerns, a body of empirical evidence to guide the management of human and organizational concerns is steadily and rapidly growing. This paradox yields several important questions: Why do managers fail to use this evidence to guide their practice? If difficulties in implementation are so widespread, why is managerial receptiveness so low to the research specifically designed to reduce impediments to implementation?

Two major reasons for failure to use these research findings are the anthropocentric focus of most research and the failure to disseminate more broadly what is known about the management of FFA. Although diffusion of the findings is important, we will not discuss it here because it has been discussed adequately elsewhere (Davis, 1986b). Instead we will focus on the anthropocentric bias of current research because we believe it to be extremely important and have not yet seen widespread discussion of this feature.

Failure to use FFA research findings is largely a result of the research perspectives most often taken. All three perspectives discussed above—technological deterministic, technological emergent, and interactive deterministic—pay little attention to the technology itself. They devote significantly less effort to understanding the technology and specific technical choices than to understanding the human issues and choices. Studies conducted from the perspective of technological determinism assume the technology is a fixed phenomenon, a given; no changes in the technology are considered (Markus & Robey, 1988). As a result, generalizations are made about the human implications of FFA, or of major categories of FFA (such as CAD or CAM), without considering the impact of different forms of the technology itself. Although the interactive determinist perspective considers the malleability of technology, little attention has been given to how different technical decisions are made and the human implications of such decisions. In management practice, human users are presumed to be more malleable than the technology; the common view is that people can change but the technology cannot.

The lack of attention devoted to specific characteristics of the technology and the technology planning process creates an impediment to the use of research findings by managers for several reasons. First, because there are so many different types of FFA, conclusions about the human needs related to the general category of FFA technology are greeted skeptically by practitioners. Most managers believe, right or wrong, that their own installation is special due to the perceived uniqueness of the particular technological configuration and the organizational context surrounding its implementation. Second, managers, such as engineers, usually have too little understanding of the human behavior associated with the use of FFA to judge which findings are most applicable to their setting. Third, research conclusions are rarely prescriptive in nature, providing little guidance for predicting costs and consequences of different technical decisions. Too often those reporting research findings write only for their scientific peers and not for managers. It is not so surprising therefore that managers seldom use research findings to guide their practice when installing FFA.

An Alternative Approach to Research and Practice

We believe that overcoming the limitations of existing approaches requires researchers and managers to examine not just the human aspects of FFA; they must also study the match of the human issues to specific characteristics of the technology. By studying this match, researchers and managers can determine which specific technical characteristics of the technology can be changed to enhance human performance. In doing so, research results can become specific and prescriptive, thereby removing the impediments to their use in practice.

To illustrate how the proposed focus on the technical-human match differs from existing approaches, note that within the technological perspective, a researcher might focus on the skills of CNC operators as a function of whether or not they program the CNC machine. Within this alternative approach, however, the *interaction* among operator, programmer, and CNC machine would be the focus. A study would try to determine how the CNC machine could be designed to allow operators to do the type of programming they need to do in order to control the production process, and allow programmers to do the level of programming they need to do in order to control the programming plan for the factory. Interaction is the key idea here. Operators and programmers make both individual and joint contributions to the success of the CNC machine output. Such a study might identify a new set of recommendations for designing CNC software, a new set of recommendations about what information should be shared between the CNC tape and the central data base of the tapes for the enterprise, and new recommendations for designing jobs to enhance the joint and individual contribution of workers.

This emphasis on technology design does not mean we should ignore the substantial contributions of humans to the successful use of FFA. We encourage a joint emphasis on both—technology and people must fit together to ensure success. The technical side is emphasized here because most social scientists fail to consider its influence, despite its importance to managers who must use the technology. Furthermore, many others already have adequately discussed the means for changing the social system to fit the technical system (e.g., L.E. Davis, 1982; Pasmore & Sherwood, 1978).

Attention to the joint influence of people and technology is not new. The Tavistock Institute for Human Relations pioneered this *socio-technical systems approach* shortly after World War II (see discussions by Emery, 1978; Trist, 1981). Sociotechnical systems (STS) theorists assume that all organizations are comprised of both technical and social systems, and that effective organizations "jointly optimize" the functioning of both systems. L. E. Davis (1982) defined a technical system as "a set of artifacts, tools, machines, facilities, methods, programs, and procedures which are the means by which the people in the organization transform the inputs [into outputs such as products or services]" (p. 2.1.7). The purpose of the technical system is to transform inputs into outputs effectively and efficiently—to change raw materials and knowledge into products and services. A social system is defined as "the set of members of the organization acting in their roles, the set of roles and their role relationships, the authority structure, communication structure, adaptation mechanisms, learning mechanisms, social system maintenance mechanisms, career structure, and so on" (Davis, p. 2.1.7). The purpose of the social system is to meet the collective needs of individuals for utilization of skills, motivating jobs, learning potential, and job satisfaction. Joint optimization of the social and technical subsystems occurs when each subsystem supports the needs of the other; unless people are sufficiently motivated, they will perform sub-optimally or leave the organization; unless inputs are efficiently and effectively transformed, products and services will be expensive and/or shoddy in quality. In either case, the organization will be less successful.

The need to fit technology to other features of the organization is a common theme in organizational theory and organizational psychology as well. Theorists long have postulated the need for organizations to achieve congruence between critical subsystems in the organization (e.g., Katz & Kahn, 1978; Thompson, 1967). Nadler and Tushman (1984) provided a congruence model of organizational design in which people and technology are distinguishable elements of a system, and these elements must be compatible with each other in order for the organization to prosper. Leavitt (1965) also identified the need for a fit between technology and people in any organizational change effort. When subsystems are out of balance, a disequilibrium occurs, resulting in the organization's reduced ability to adapt to a changing environment. STS and open systems theorists focus on the fit between the technical and social subsystems, emphasizing the need to establish congruence and balance in order to ensure adaptability of the organization.

Sociotechnical Match

Although this notion of a sociotechnical fit has existed for some time, it has only recently guided research and practice concerned with FFA. Vague concepts, such as social system, technical system, and joint optimization or congruence, do not easily lend themselves to precise and systematic research and practice. Nevertheless, because the perspective provides a set of concepts with which both technical and social issues can be considered simultaneously in any study of FFA, it is attractive and becoming more popular. In fact, it is almost impossible to join in conversations with technological managers today without their mentioning the need to match the human and technical elements of an organization.

One of the reasons for the increasing popularity of this alternative approach is that technology and people are viewed to be *equally* malleable. When they plan change, managers adopting this perspective elaborate technical choices and consequences in as much detail as human choices and consequences. Consequently, managers come to a better understanding of the constraints imposed on the human organization by certain technology choices. With this knowledge about implications of specific technical choices, managers can modify such choices early in their planning, before costly or damaging consequences of inappropriate choices are felt.

The STS perspective is not without its flaws. Although STS provides many useful concepts, operationalization of these ideas is weak and incomplete. In order to guide research and practice better, the STS perspective must provide better definitions of its critical concepts. In particular, better definition is needed for the idea of "fit" or "match" between the technological and social systems.

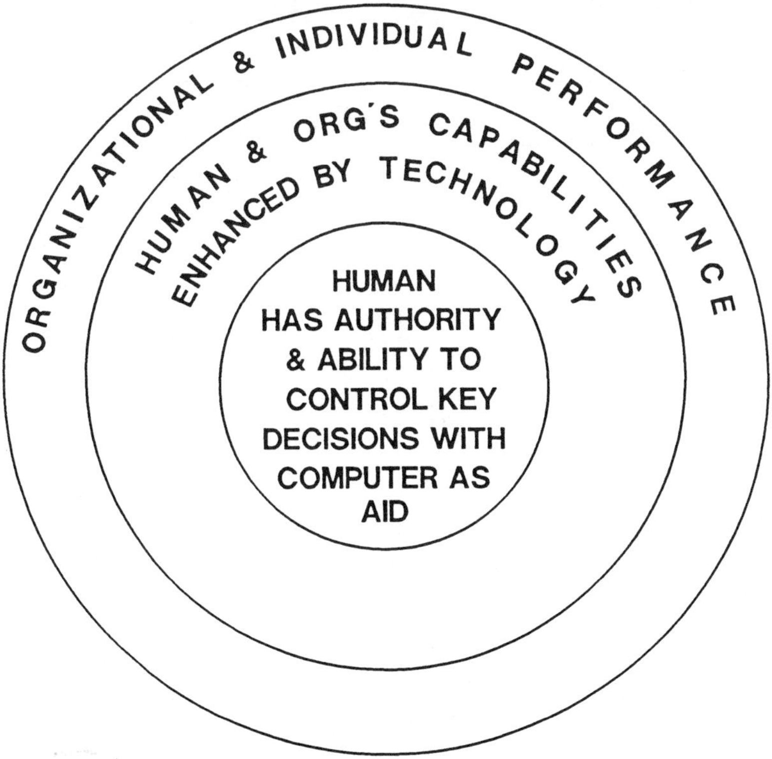

Figure 2.2. Depiction of Sociotechnical Match

We propose such a definition (see Figure 2.2) specifically as it pertains to FFA. The figure depicts the match between the technological and social subsystem as defined at three levels of generality. At the most general level, sociotechnical match is based on whether the social and technical subsystems enhance organizational and individual performance; thus match is defined by its effects on performance (Blumberg, 1988; Trist, 1981). Organizational performance may be indicated by productivity, profitability, quality and quantity of production, and human resource outcomes, such as turnover, injuries, and absenteeism.

Since such a general definition of match between the social and technical subsystems has only limited use for research and practice, we propose a more specific version. Perhaps the most universally accepted assumption among STS proponents is that a match is achieved when the computerized technology has been designed to enhance the capabilities of the human and the organization rather than simply to replace humans or to perform tasks that they perform poorly (Cherns, 1976; Trist, 1981). Moreover, it is assumed that full gains in human and organizational performance will be achieved only when the human enhancement potential of technology is explicitly considered in the FFA planning process. For this reason, the enhancement potential of technology is a critical element within the larger circle of performance in Figure 2.2.

To illustrate this more specific definition of match, consider a welding robot that is designed only to replace welders. The use of a robot in this manner does not meet the requirements for a sociotechnical match because, at the very least, the social needs of welders and the welding group are ignored; in the worst case such needs may be severely damaged. Long-time friends may no longer be able to interact and share information; work may now be done in isolation. Also, if there was inadequate training and preparation for the change, workers may blame themselves for failure to use the new equipment optimally rather than attributing these failures to inadequate management of the implementation process. As a result, job satisfaction and organizational commitment may decrease, absenteeism and turnover may increase, and substandard performance is likely. In contrast, a welding robot that removes the tedious and physically demanding work of the welder, while being supervised by welders is far closer to a sociotechnical match. Here the work of welders is augmented, not mechanized out of existence.

Although defining "match" by the extent to which the technology enhances the potential of human and organizational capabilities provides greater specificity, researchers and managers concerned with individual behavior require additional clarity about which capabilities need to be enhanced. Debate in the literature continues about this issue, but many STS researchers would agree that human capabilities are enhanced when the human retains authority and the ability to control key decisions about the tasks to be accomplished (this is the STS notion of self-regulation of fluctuations and disruptions; see Schilling, Schuepbach, & Ulich, 1988; Trist, 1981). Studies of human-machine interactions indicate that such interactions have more positive effects on individuals' psychophysiological states of fatigue, monotony, and satiation when the individuals possess greater control over elements of the technical system, such as how tasks will be performed or exploring for alternative paths (Hacker, 1986).

For humans to retain control, computers must serve them and aid their efforts, and this is the most specific aspect of "match" shown in Figure 2.2. Corbett (1985, 1988) describes a human-centered approach to human-machine system design. Such an approach rejects the more traditional view of human-machine system design in which machines control all functions that the technological state of the art permits, while humans control those functions that do not yet lend themselves to technical solution. Consistent with our definition of sociotechnical match as technology that enhances human capability, Corbett (1988) describes the aim of the human-centered approach as the utilization of existing skills and the development of new skills. To achieve this goal, he proposes to increase operator control of the technical system. The operator should have knowledge about how the computer controls production and the authority to alter this control when necessary. The operator should have the freedom to change operations when needed.

In sum, sociotechnical match can be defined as human control over the tasks that must be accomplished, the enhancement of human capabilities by the technology, and the enhancement of individual and organizational performance.

This definition of sociotechnical match helps to identify three concerns that must be addressed by those interested in the human side of FFA: the identification of matches and mismatches, appropriate matches for different levels of analysis, and facilitating matches in organizations. These concerns require additional research and do not currently yield obvious recommendations for practice. With more research attention, however, this situation is likely to shift toward more prescriptive advice. These concerns are not independent nor additive. All must be addressed to ensure the successful implementation and management of FFA, although the relative importance of each is unknown. We believe that the probability of successful FFA implementation is increased to the degree that all are considered.

Identification of Matches and Mismatches

This concern is an ultimate objective of STS-oriented research and practice involving flexible factory automation. Existing and potential matches and mismatches must be identified. To identify when a mismatch *currently* exists, the manager needs to know which social and technical factors to examine and how to interpret the information obtained about each factor. To identify when a mismatch is *likely* to occur, the manager needs a predictive model to point out factors demanding attention. STS research must progress substantially before this objective can be achieved.

Appropriate Matches for Different Levels of Analysis

The second concern is to identify specific types of matches that exist at different levels of analysis. Matches can exist at the most micro-level of the human-machine interaction, such as when dials, menus, and information displays are designed to enhance the operator's comprehension and knowledge of relevant information. Matches can also exist at a middle level of job design, such as when teamwork is introduced to promote use of the flexibility and variety provided by certain kinds of FFA. Finally, matches can exist at a macro-level of organizational design such as when the formalization of procedures and decentralization of decision-making authority are matched to the needs of specific types of FFA (e.g., when the authority to decide to shut down an assembly process is delegated to those in the organization who are closest to the assembly). Consequently, critical features of the technology at different levels of analysis must be matched with critical features of the social system at the same level of analysis. Ideally, these critical technical features should be of sufficient generality to pertain to a reasonably broad group of technologies, yet also of sufficient specificity to make clear as to what parts of the technologies they refer. When judged against these criteria for the identification of technical features, recommendations concerning technology offered by organizational theorists and designers who examine relationships with general variables, such as task variability (i.e., number of exceptions encountered when doing work), are often too diffuse to help the researcher or manager adhering to an STS perspective.

Facilitating Matches in Organizations

The third concern is to understand how sociotechnical matches can be facilitated in the technology planning process. Perrow (1983) discussed this question in regard to integrating human factor considerations into the design of such complex technological systems as nuclear power plants and missile systems. Similarly for FFA systems, it is important to develop mechanisms that will encourage technology planners to consider sociotechnical matches. The creation of simulations that include trade-offs for technical decisions with different human-related implications could be one such mechanism; efficient forms of user participation could be another. Information on the most effective mechanisms for facilitating sociotechnical matches in different organizational settings is greatly needed.

Research Examining Sociotechnical Matches and Mismatches

Of the three concerns discussed above, the identification of matches and mismatches has received the most research attention; therefore, it will be discussed in more detail. To illustrate how research on sociotechnical matches is done, STS approaches to the study of CAD, human-computer interaction, software development, and flexible manufacturing cells will be described.

Computer-Aided Design

Sinclair and his colleagues (Sinclair, Siemieniuch, & John, 1989) used an STS perspective to develop computer-aided design (CAD) technology. Rather than simply accepting the technology provided by vendors (as in the technological determinism approach), they observed the work of R&D engineers and interviewed them to learn how the product design process evolves. They then used this information to change CAD technology so that it could enhance the design process. Furthermore, they wanted the software to be capable of being adapted to the evolution of the organization.

Sinclair et al.'s results reveal many individual differences in the way that R&D engineers approach the product design process. For example, some designers prefer visual displays, while others prefer specific design rules to follow. The authors also found that R&D engineers spend considerable time during the initial stages of product design in discussions with colleagues and in reviewing data bases to uncover potential implications of their designs, such as tolerances needed and product materials that could be used.

Sinclair et al. drew several conclusions about sociotechnical matches for CAD technology. First, CAD technology must be designed to be "friendly" to a variety of users, especially early in the design process when so many potential user groups are providing input into the design. To be friendly to multiple users implies that the software needs to be sufficiently flexible to allow for different levels of facility with computers, different uses of language to correspond to each group's nomenclature, and different levels of detail in accessing the design file to correspond to each group's knowledge and interest in the design.

Second, CAD software must allow easy manipulation of the design data early in the design process in order to allow easy experimentation with different design ideas. This could involve the creation of small, dedicated files of design-relevant information concerning tolerances, stresses, materials, manufacturing processes, marketability, and other factors. The CAD software could also query users about the type of information they would find useful and then present them with samples of the requested information.

Third, CAD software should allow users to maintain their individual differences of approach in the design process while enhancing their capabilities. To accomplish this, the software should provide the R&D engineer with the capability to make visual manipulations of ideas (not just using programming or formulas), with flexibility in use of software (such as having a detailed menu as well as the opportunity to bypass it), and with preprogrammed design rules that can be overridden easily by the CAD user.

Fourth, the CAD system should support multidisciplinary teams of designers as well as suppliers, subcontractors, and clients. Close contact should be maintained to allow especially free communication and exchange of ideas. Note that this makes obsolete the traditional notion of an organizational boundary. With FFA, it will be necessary to manage networks of organizations, not just a single team of designers within the manager's own organization.

The CAD software proposed by Sinclair et al. (1989) does not exist yet. Nevertheless, these recommendations are specific enough to be used, and they will improve the sociotechnical match because human capabilities will be enhanced by these changes in CAD technology. Rather than accepting the technology as given, managers can require vendors to change the technology design in order to enhance the capability of their engineers to use the technology. This request should meet little objection from those CAD vendors who are sensitive and responsive to customer needs.

Human-Computer Interaction

A second example of an STS approach to the study of human issues in FFA comes from Rouse and his colleagues, who have conducted a series of studies (mostly laboratory experiences and simulations) involving USAF pilots in order to design flight control technology (e.g., Rouse, 1988; Rouse, Geddes, & Curry, 1987; Rouse, Rouse, & Hammer, 1982). Rouse proposes a "support system concept of task allocation" (or "computer-aiding"), in which automation is a backup to be used either when anticipated operator performance is unacceptable or when the operator chooses for some reason to relinquish control. The aim is to preserve the operator's discretion to act in appropriately innovative ways; thus the operator can remain in control while the computer-based system provides support that helps the operator to avoid the negative consequences of inappropriate actions. Rouse et al. (1987, p. 91) state that "the question is not *when* should humans be in charge, but *how* can designers provide means for overcoming humans' limitations and enhancing humans' abilities so as to extend the range within which humans can remain in control of the technologies underlying their systems?" (emphasis in original).

A study by Rouse, Rouse, and Hammer (1982) demonstrates this approach. These investigators designed an onboard computer-based information system to support the performance of two-man flight crews under situations of increasing stress. The computer was used as an aid for retrieval and display of information, as well as for monitoring the use of information concerning normal, emergency, and double emergency situations. With the help of the computer system, pilots were able to make decisions and execute procedures with fewer errors. Instead of automating actions to redress emergency situations, this system augments the abilities of human pilots to make critical decisions.

From the research of Rouse and his colleagues, principles of human-computer task allocation have been identified that overcome the problems of the traditional automation philosophy of task allocation. These principles include the following:

- Task allocation should occur dynamically as the task proceeds (i.e., the human or the computer may perform the same task over time in response to differing needs of the situation).
- Tasks should be categorized first by who could potentially perform them (human, computer, or both).
- For tasks that would be performed by the human or the computer, the task should be allocated to the computer only when (a) the operator prefers it, (b) the operator does not intend to perform the task, (c) the operator cannot perform the task acceptably, (d) the operator's performance on other tasks will be degraded, or (e) the operator is unaware that the task needs to be performed.
- For tasks not allocated to the computer at any given time, the computer can either perform part of the task or make the task easier, such as by changing the form of the display.
- If operator performance begins to decrease gradually, the computer can gradually take over the operator's tasks.

In addition to these principles of task allocation, Rouse et al. (1987) also described software that would help implement these principles. This software includes a program that estimates current and predicted states of the operator; for instance, it uses the operator's goals, plans, and current activities to estimate the operator's awareness, intentions, resource utilization, and future performance. Also included is a program that provides a means of overcoming natural human limitations by detecting anomalies, classifying errors by causes, and identifying error correction possibilities. Finally, the software includes a program that manages the efficient flow of information between the operator and the computer, such as by prioritizing messages or selecting modalities of presentation to fit with operator desires. It is apparent that sociotechnical matches can take the form of software that allows for dynamic task allocation between the computer and the human by "reading" the needs of the human and helping the human to perform more efficiently and effectively.

Software Development

A third example of research illustrating the STS perspective examined the process by which software is developed (Bendifallah & Scacchi, 1988; Garg & Scacchi, 1989; Scacchi, 1984). Scacchi and his colleagues have conducted a series of studies involving computer systems analysts to understand how technology can be designed to facilitate the software development process.

Scacchi believes that the software development process is essentially a cooperative group effort that requires software users to work with software developers; technology that facilitates this cooperative effort is needed. Although some computer programs that foster cooperative work exist today, most support a particular division of labor that must be passively accepted by the group. Such predefined divisions of labor are not consistent with STS goals. Scacchi has developed software that helps the group create its own effective work structure during its developmental work.

Scacchi's technology enhances human capabilities in several ways. First, it helps the group organize its work to follow more closely the typical sequence of procedures in the development process. Second, it helps allocate responsibilities to group members based on their respective skills. Third, it compares group activities with the planned sequence of tasks and provides feedback to the group about what has been accomplished, likely consequences of pursuing the particular path chosen, and alternative paths to select. For Scacchi, an optimal match between the social and technical subsystems occurs when the technology facilitates the creation of an effective and efficient work structure. This type of fit is particularly needed in work units where role responsibilities and decision making are decentralized, shared, and participative in contrast to units with more traditional bureaucratic forms having a highly fragmented division of labor, close supervision, and narrow spans of control.

Flexible Manufacturing Cells

A final example illustrating the STS approach that we advocate comes from work in progress by Majchrzak and her colleagues. This work aims to develop a computer simulation model for predicting the human and organizational infrastructure likely to be needed if any particular flexible manufacturing cells (FMC) design is implemented in any particular factory setting (Majchrzak, 1988b; Majchrzak, in press). The intent of this research (supported by Digital Equipment Corporation) is to provide FMC designers with early simulations of the

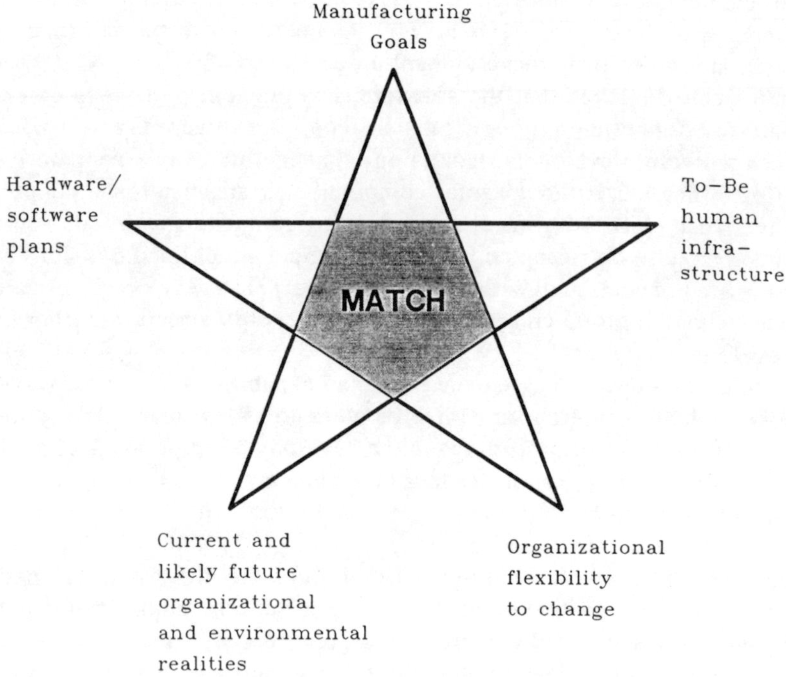

Figure 2.3. Elements of a Sociotechnical Match for FMCs

human consequences of their designs so that changes to the designs can be made before any action is taken to implement the design. In essence, this research is focused on how to create a sociotechnical match.

To guide this research, five elements of a sociotechnical match, depicted in Figure 2.3, have been identified and elaborated. The first element is the "To-Be" human infrastructure that will need to be implemented when the cell technology arrives. The "To-Be" human infrastructure consists of the sets of human tasks, work design, skill requirements, training and selection programs, formal and informal structural arrangements, and the performance management system (e.g., pay). The second element is management's strategic goals for the cell, which can potentially include priorities for flexibility, cost, quality, and innovation. A third element is the current and likely future realities of the organization and its environment. These realities can also be thought of as constraints or "givens." They include the critical technical variations in the production environment (such as "hot jobs," small batch sizes, or short work-cycle times), resource dependencies in the environment (such as union contractual arrangements, customer contracts, and supplier agreements about quality expectations of incoming materials), and management philosophies (such as management priorities concerning the trade-off between efficiency and human effectiveness, or between short-term and long-term horizons). A fourth element of the sociotechnical match is the cell technology itself: its hardware/software capabilities. Within this element, features include reliability, accuracy, back-up alternatives during breakdowns, and electronic interconnectedness between machines. The final element of the match is the organization's flexibility to adopt the changes needed to achieve a match. Flexibility for change involves both the ability and willingness of individuals to change, as well as the structural flexibility (e.g., an organic rather than bureaucratic structure) of the organization to tolerate and adapt to change.

For these five elements to be useful in guiding the development of a predictive simulation model, each element must first be carefully defined in terms of specific variables. After achieving consensus on variables and definitions, the critical task of specifying those interrelationships among the elements that yield optimal performance can proceed. Thus far in our research, we have identified matches among the five elements as they pertain to one aspect of the To-Be human infrastructure: human tasks that are likely to be needed if the FMC is to process parts effectively and efficiently. Our tentative conclusions from this work include the following:

1. Achieve consensus on the boundaries of a focus; don't assume such a consensus exists. Moreover, once such a consensus is established, constantly refer back to it since it is exceedingly easy to ignore those boundaries; doing so will yield research and practice that are not well specified and are thus difficult to grasp. In our project, we found we needed to achieve consensus (among manufacturing engineers, plant managers, human resource professionals, and our research team) about (a) the technologies involved in an FMC, (b) the industry and plant settings to which our FMC applied, (c) how an FMC was distinguished from "cellular technology" as used in practice today, (d) what operations and human functions could be considered within the FMC and which were clearly beyond our boundaries (e.g., was product engineering considered within the cell for the purpose of the project?),

and (e) how permeable the boundaries between the cell and its larger factory environment would be (e.g., what inputs from the factory into the cell and what outputs to the factory would be considered and how?). Achieving this consensus has been essential not only for keeping practitioners interested in the research, but for keeping our research focus clear.

2. No sociotechnical research project is without a set of value-driven assumptions (e.g., about the role of humans or the likely capabilities of technology despite vendor promises). Since an STS project—as we have described it in this chapter—is focused on identifying different matches under different situations, these value-driven assumptions must be explicitly considered as another influential variable in determining which matches will work for the organization and which will fail. In our research project, we have found it necessary to specify two types of assumptions: those that we think are so critical to the performance of the cell that we will not allow them to be overridden by the FMC designer using the simulation, and assumptions that we might prefer a designer to make but will allow to be overridden. An example of the former is the assumption that humans within the cell should perform tasks that allow them to make improvements in the human/machine system of the cell; without some degree of continuous improvement in the cell, organizational competitiveness will deteriorate. An example of an assumption that we will allow designers to override (although we would prefer that they not do so) is the involvement of cell personnel in scheduling work orders—if management wants to keep this involvement and responsibility out of the hands of cell personnel, we have decided that our simulation must still allow for an assessment of the human infrastructure needs given that scenario.

3. A substantial amount of knowledge about the typical production environment of an FMC and the hardware/software capabilities of FMCs is essential in order to make predictions about human tasks needed under different production/capabilities combinations. Unfortunately, much of this detailed knowledge is not in the empirical literature on human aspects of FFA because of the aggregate level at which FFA systems are described. As a result, many plant visits and Delphi interview studies with manufacturing engineers have been needed to match the realities of practice against theoretical predictions. This iterative process has slowed the research progress but has been invaluable in ensuring predictions with practical significance.

4. Given the requisite knowledge, generating predictions about matches among the five elements depicted in Figure 2.3 has proven to be not nearly as difficult a task as originally anticipated. Hundreds of predictions have been generated thus far. The almost insurmountable challenges appear to be: how to determine which of these predictions are *the most critical* to organizational performance, how to summarize or "meta-analyze" these numerous predictions into a parsimonious set of a few key propositions, and how to present this very complicated model to colleagues and practitioners to solicit input and determine face validity. These are challenges that we still face.

Summary

These four applications illustrate how research on human aspects of FFA can be conducted within an STS perspective. Technology is considered as malleable as the social system. As a result, specific recommendations for redesigning technology can be made. Finally, redesigns of technology can help to enhance human performance, not simply complement or replace it. Only when researchers and managers pay sufficient attention to the human side of FFA and use this technology to enhance human and organizational performance will we achieve the promise of CIM.

Relation of the STS Approach to Social Psychology

The STS approach to research on the human side of FFA suggests a focus on technology issues as well as social, human, and organizational issues. Social psychologists have traditionally limited their research attention to individual and group behavior, leaving a serious examination of technological factors to the computer scientist or engineer. We have argued that such a limited perspective unnecessarily constrains the impact of social psychology. Yet, how can a focus on technology fit with the social psychologist's traditional exclusive focus on the social system?

One way of realigning the current interests of social psychologists toward an STS approach is to reconsider the domain of knowledge covered by social psychology. Sherif and Sherif (1956) have defined social psychology as the study of the behavior of individuals in relation to social stimulus situations. It is apparent from the research on human impacts of FFA that the implementation of FFA can potentially create many changes to the human/social/organizational work environment; thus it is clearly a social stimulus situation. Moreover, social psychology also studies the types of variations in social stimuli that will elicit certain desired behaviors. Consequently, social psychology would include the study of variations in FFA that facilitate desired behavioral responses.

However, in suggesting that social psychology can include the study of the technology aspects of FFA, we must not overlook the fact that most social psychologists have little training in engineering or computer science. Moreover, most social psychology journals do not typically publish articles that describe technology-related issues in detail. The dilemma for the social psychologist, then, is how to conduct research at the fringes of traditional social psychology with little formal training on the issues needing the greatest attention. One strategy for resolving this dilemma is to team with an "enlightened" engineer or computer scientist, and to tailor articles differently depending on the audience. The technically oriented audience will prefer more discussion of the technology, the social psychology audience will prefer more discussion of the human issues, and the practitioner audience will prefer a discussion integrating the two outlooks. This strategy is often adopted in Europe, as illustrated by such recent research programs as the United Kingdom's Alvey Research Program or the European Strategic Program of R&D in Information Technology (ESPRIT).

An alternative strategy is for researchers to cross-train in the particular types of technologies of interest. This cross-training need not involve a complete engineering degree; rather, courses on FFA often can be taken in engineering or computer science departments, or enough practicing technologists can be interviewed as informants to gain an understanding of the critical technical decisions that are made and why. Since forging intellectual partnerships is often very difficult because of conflicting personal interests and freedom needs, this alternative strategy may be more feasible. However, the lone social psychological researcher will still have the problem of where and how to publish in professional journals. Such a researcher will most likely need to de-emphasize the technology issues in social psychology articles, saving the more technological discussion for human factors and practitioner-oriented journals. Though such interdisciplinary work is often difficult, social psychologists interested in the STS approach to studying human issues involved in FFA should take comfort from knowing that other social psychologists are successfully studying these topics and that this area of research is important for the future of our industrial society.

References

Adler, P. S. (1989). CAD/CAM: Managerial and organizational challenges of integrating CAD and CAM. *IEEE Transactions in Engineering, 36*, 202-215.

Argote, L., Goodman, P. S., & Schkade, D. (1983). The human side of robotics: How workers react to a robot. *Sloan Management Review, 24*(3), 31-41.

Ayres, R. U., & Miller, S. M. (1983). *Robotics: Application and social applications.* Cambridge, MA: Harper & Row.

Bendifallah, S., & Scacchi, W. (1988). *Work structures and shifts: An empirical analysis of software specification teamwork* (Tech. Rep.). Los Angeles: University of Southern California, Computer Science Department.

Blumberg, M. (1988). Towards a new theory of job design. In W. Karwowski (Ed.), *Ergonomics of hybrid automated systems I* (pp. 53-59). Amsterdam: Elsevier.

Cherns, A. (1976). The principles of sociotechnical design. *Human Relations, 29*, 783-792.

Corbett, J. M. (1985). Prospective work design of a human-centered CNC lathe. *Behavior and Information Technology, 4*, 201-214.

Corbett, J. M. (1988). Designing hybrid automated systems: A European perspective. In W. Karwowski (Ed.), *Ergonomics of hybrid automated systems I* (pp. 167-172). Amsterdam: Elsevier.

Davis, D. D. (1986a). Designing organizations for technological innovation, productivity, and quality of work life: A human resource perspective. In D. O. Gray, T. Soloman, & W. Hetzner (Eds.), *Technological innovation: Strategies for a new partnership* (pp. 145-170). Amsterdam: Elsevier.

Davis, D. D. (Ed.). (1986b). *Managing technological innovation: Organizational strategies for implementing advanced manufacturing technologies.* San Francisco: Jossey-Bass.

Davis, D. D. (1986c). Integrating technological, manufacturing, marketing, and human resource strategies. In D. D. Davis (Ed.), *Managing technological innovation: Organizational strategies for implementing advanced manufacturing technologies* (pp. 256-290). San Francisco: Jossey-Bass.

Davis, D. D., Roitman, D., Carlopio, J., & Fleischer, M. (1987, August). *Technological change in the workplace: Labor and management issues.* Paper presented at the meeting of the American Psychological Association, New York.

Davis, L. E. (1982). Organization design. In G. Salvendy (Ed.), *Handbook of industrial engineering* (pp. 2.1.1-2.1.9). New York: John Wiley.

Dean, J. W. (1987). *Deciding to innovate: How firms justify advanced technology*. Cambridge, MA: Ballinger.

Emery, F. (1978). *The emergence of a new paradigm of work* (Tech. Rep.). Canberra: Australian National University, Centre for Continuing Education.

Ettlie, J. E. (1986). Facing the factory of the future. In D. D. Davis (Ed.), *Managing technological innovation: Organizational strategies for implementing advanced manufacturing technologies* (pp. 72-104). San Francisco: Jossey-Bass.

Ettlie, J. E. (1988). *Taking charge of manufacturing*. San Francisco: Jossey-Bass.

Garg, P. K., & Scacchi, W. (1989). ISHYS: Designing an intelligent software hypertext system. *IEEE Expert, 4*(3), 52-63.

Gasser, L. (1986). The integration of computing and routine work. *ACM Transactions of Office Information Systems, 4*, 205-225.

Hacker, W. (1986). Complete vs. incomplete working tasks: A concept and its verification. In G. Debus & H. W. Schroeff (Eds.), *The psychology of work and organization* (pp. 23-36). Amsterdam: Elsevier.

Hunt, H. A., & Hunt, T. L. (1983). *Human resource implications of robotics*. Kalamazoo, MI: Upjohn Institute.

Jaikumar, R. (1986). Post-industrial manufacturing. *Harvard Business Review, 6*, 69-76.

Jelinek, M., & Goldhar, J. D. (1986). Maximizing strategic opportunities in implementing advanced manufacturing systems. In D. D. Davis (Ed.), *Managing technological innovation: Organizational strategies for implementing advanced manufacturing technologies* (pp. 220-238). San Francisco: Jossey-Bass.

Katz, D., & Kahn, R. L. (1978). *The social psychology of organizations* (2nd ed.). New York: John Wiley.

Kling, R., & Scacchi, W. (1982). The web of computing: Computer technology as a social organization. *Advances in Computers, 21*, 2-90.

Leavitt, H. J. (1965). Applied organization change in industry. In J. G. March (Ed.), *Handbook of organizations* (pp. 1144-1170). Chicago: Rand McNally.

Majchrzak, A. (1988a). *The human side of factory automation*. San Francisco: Jossey-Bass.

Majchrzak, A. (1988b, August). *Towards a framework for identifying organizationally-compatible manufacturing technology*. Paper presented at the International Conference on Ergonomics of Advanced Manufacturing Technology and Hybrid Systems, Louisville, KY.

Majchrzak, A. (in press). Effect of CAD on the jobs of drafters and engineers: A quantitative case study. *International Journal of Man-Machine Studies*.

Majchrzak, A., Mosher, P., & John, R. (1988). *Technological change and structural contingency theory*. Unpublished manuscript, University of Southern California, Department of Human Factors, Los Angeles.

Majchrzak, A., Nieva, V., & Newman, P. (1986). Adoption and use of computerized manufacturing technology: A national survey. In D. D. Davis (Ed.), *Managing technological innovation: Organizational strategies for implementing advanced manufacturing technologies* (pp. 105-126). San Francisco: Jossey-Bass.

Manufacturing Studies Board. (1986). *Human resource practices for implementing advanced manufacturing technology*. Washington, DC: National Academy Press.

Markus, M. L., & Robey, D. (1988). Information technology and organizational change: Causal structure in theory and research. *Management Science, 34*, 583-597.

Mohr, L. B. (1987). Innovation theory: An assessment from the vantage point of new electronic technology in organizations. In J. M. Pennings & A. Buittendam (Eds.), *New technology as organizational innovation: The development and diffusion of microelectronics* (pp. 13-31). Cambridge, MA: Ballinger.

Nadler, D. A., & Tushman, M. L. (1984). A congruence model for diagnosing organizational behavior. In D. A. Kolb, I. M. Rubin, & J. M. McIntyre (Eds.), *Organizational psychology* (4th ed., pp. 586-603). New York: Prentice-Hall.

Office of Technology Assessment. (1984). *Computerized manufacturing automation.* Washington, DC: Author.

Parsons, H. M. (1988). Human factors in robot design and robotics. In D. J. O'Borne (Ed.), *International review of ergonomics* (2nd ed., pp. 151-176). London: Taylor & Francis.

Pasmore, W. A., & Sherwood, J. J. (Eds.). (1978). *Sociotechnical systems: A sourcebook.* San Diego: University Associates.

Perrow, C. (1983). The organizational context of human factors engineering. *Administrative Science Quarterly, 23,* 521-541.

Roitman, D. B., Liker, J. K., & Roskies, E. (1988). Birthing a factory of the future: When is "all at once" too much? In R. H. Kilmann, T. J. Covin, & Associates (Eds.), *Corporate transformation: Revitalizing organizations for a competitive world.* (pp. 205-246). San Francisco: Jossey-Bass.

Rouse, S. H., Rouse, W. B., & Hammer, J. M. (1982). Design and evaluation of an onboard computer-based information system for aircraft. *IEEE Transactions on Systems, Man, and Cybernetics, SMC-12,* 451-463.

Rouse, W. B. (1988). Adaptive aiding for human/computer control. *Human Factors, 30,* 431-443.

Rouse, W. B., Geddes, N. D., & Curry, R. E. (1987). An architecture for intelligent interfaces. *Human-Computer Interaction, 3,* 87-122.

Scacchi, W. (1984). Managing software engineering projects: A social analysis. *IEEE Transactions of Software Engineering, SE-10,* 45-59.

Schilling, A., Schuepbach, H., & Ulich, E. (1988). Sociotechnical design of advanced manufacturing systems. In W. Karwowski (Ed.), *Ergonomics of hybrid automated systems I* (pp. 77-78). Amsterdam: Elsevier.

Sherif, M., & Sherif, C. W. (1956). *An outline of social psychology.* New York: Harper.

Sinclair, M. A., Siemieniuch, C. E., & John, P. A. (1989). A user-centered approach to define high-level requirements for next-generation CAD systems for mechanical engineering. *IEEE Transactions on Engineering Management, 36*(4).

Thompson, J. D. (1967). *Organizations in action.* New York: McGraw-Hill.

Trist, E. L. (1981, June). *The evolution of sociotechnical systems: A conceptual framework and an action research program* (Occasional Paper No. 2). Toronto: Ontario Quality of Work Life Centre.

Waterbury, R. (1986, February). IBM: Management meets the automation challenge. *Assembly Engineering, 29*(2), 18-20.

Works, M. T. (1987). Cost justification and new technology: Addressing management's "no!" to the funding of CIM. In L. Bertrain & L. Hale (Eds.), *A program guide for CIM implementation* (2nd ed., pp. 172-180). Dearborn, MI: Society of Manufacturing Engineers.

Yoon, S. J. (1988). *An exploratory study of the relationship between AMT and organizational structure.* Unpublished doctoral dissertation, School of Business Administration, Pennsylvania State University.

Zuboff, S. (1988). *In the age of the smart machine.* New York: Basic Books.

3

Factory Automation and Production Jobs: A Longitudinal Study

VERONICA F. NIEVA
PAUL D. NEWMAN
GEZA BOTTLIK

The recent introduction of computerized manufacturing technology has revived many of the controversies stimulated by earlier developments in automation and generated new expectations and concerns. These new manufacturing technologies include a wide variety of programmable tools used in various production and assembly processes. Examples on the factory floor include robots, computerized numerically controlled (CNC) machines, computer-aided inspection and test equipment, and computerized insertion machines. The new technologies are often implemented by manufacturing companies as part of a coordinated approach to address their productivity and efficiency problems.

While these new approaches are viewed as crucial to addressing problems of productivity and competitiveness at both the company and

---

AUTHORS' NOTE: We thank the National Science Foundation and the Upjohn Institute for Employment Research for their support of this study. We also thank the employees at the factory, who gave us countless hours and many insights that form the basis of this chapter.

national levels, questions have been raised regarding whether such improvements are achieved at the expense of individual workers. Worker advocates have voiced their concerns regarding potential problems in terms of job losses, social dislocation, and general deterioration of the working environment.

This chapter examines the employment and job changes that accompanied the introduction of new manufacturing technology in a plant that manufactured printed wire board assemblies (PWBAs). As discussed in a later section of this chapter, this initiative was part of a corporate push to regain the company's former competitive position in the international market. The research reported here was part of a multimethod case study conducted between 1985 and 1987 that focused on the implementation and impact of state-of-the-art automated manufacturing technologies.

The first section of this chapter reviews the arguments and available empirical evidence concerning the potential impact of the new technologies on workers. We then present the methodology and data sources used in the study, describe the organizational setting in which the research was conducted, and discuss the innovations in technology and organization implemented at the facility during the period that was studied. These descriptive sections are followed by data on the types of job changes expected to be vulnerable to the technological and organizational innovations implemented in the PWBA plant during the period studied—i.e., changes in the *number* of jobs, job *task structure*, and job *quality*.

In general, the discussion in this chapter is based on the combination of many sources of quantitative and qualitative data obtained over the three-year duration of the study. As is often the case in real-time, on-site research, multiple developments emerged simultaneously, and an important aspect of the research task was to sort through the various sources of information collected in order to distill consistent patterns of relationships and explanations.

Background

Job Displacement Versus Job Creation

Will workers become obsolete? A central concern for many employee advocates, particularly labor union representatives, is the power of the new forms of manufacturing technologies to substitute machines for human labor, thereby resulting in extensive job displacement. Entire job classes may be eliminated and/or created as tasks are transferred from people to machines. Robots, for example, are taking over the jobs of spot welders and spray painters in automaking (Hunt & Hunt, 1983) and are moving into other occupations in other industries, such as printed circuit board assembly in electronics. Manual loading tasks are being replaced by advanced materials handling equipment.

Various scholars have predicted greater or lesser levels of worker displacement resulting from new technology implementation and related efforts to streamline manufacturing procedures. Early forecasts predicted massive displacements. For example, Ayres and Miller (1982) predicted that between 1 and 3 million production workers out of 8 million would be replaced by robots alone by the year 2002. However, other scholars, including Hunt and Hunt (1983), Lublin (1981), and Leontief and Duchin (1984), have predicted much more gradual change and lower levels of eventual displacement. Productivity improvement efforts that include the implementation of new manufacturing technologies frequently require extensive restructuring that can only be implemented incrementally (Bednarzik & Schoepfle, 1985; Hunt & Hunt, 1983). Other factors, such as the state of the economy, employer philosophies regarding the use of their human resources, and the state of the technology itself, are likely to affect the extent to which jobs and workers are displaced (Andelin, 1983; Bednarzik & Schoepfle, 1985; Levin & Rumberger, 1983).

Job Task Structure

Studies are increasingly emphasizing that subtle changes in job duties and skill requirements resulting from automation are more far-reaching than sheer issues of displacement (Hunt & Hunt, 1983). The use of programmable automation can fundamentally change the relationships between worker and machine. On a basic level, the new technology can alter the nature and mix of tasks required. In some instances, the creation or elimination of tasks can change the nature of a given job without creating new jobs or eliminating old ones. In the Westinghouse Defense and Electronics System Center in Baltimore, Maryland, for example, materials handlers learned to carry out such tasks as processing receivable materials by computerized information-processing techniques that replaced the old paper and pencil methods. Welders in the same factory learned to monitor, operate, and edit programs for the welding robot that took over some of the welding operations (Argoff, 1984).

Other jobs may be significantly changed either by "job blending" (where the jobs done by two workers in previously separate occupational categories are blended into one) or by a significant alteration in the nature of the work caused by the use of computer-aided equipment. For example, CNC (computerized numerical control) jobs in large German firms have tended to combine production operator responsibilities with programming and planning functions (Hartmann, Nicholas, Sorge, & Warner, 1984). In some cases, production jobs have also expanded to take on simple maintenance tasks as well as machine set-up operations.

Job changes may be so extensive that essentially new jobs are created. In addition, entirely new jobs, like software development and maintenance, new technology equipment manufacturing, and technical training, have been created by programmable automation, although some of those new jobs may be transitory as the new technology acquires more sophisticated capacities and implementation moves into more routinized phases.

Job Quality

Another issue regarding technology's impacts concerns indicators of job quality. Some scholars suggest that work fragmentation resulting from advances in microelectronics in the workplace will contribute to the de-skilling of jobs (Braverman, 1974; Cooley, 1983). Increasing automation, it is argued, may supplant the operator's role in skilled production tasks, leaving only the less challenging components of the job for human workers. Other predictions stem from earlier themes regarding worker alienation resulting from technology (e.g., Blauner, 1964). Such forecasts focus on programmable technology's capacity for eroding worker control because of the increasing capacity for machine-controlled pacing and monitoring (Shaiken, Herzenburg, & Kuhn, 1986). From a study of workers using CNC equipment, Shaiken and his colleagues suggested that the shift from conventional machining equipment tends to reduce the role of workers in production. This recent study corroborated earlier findings (e.g., Hazelhurst, Bradbury, & Corlett, 1969) that demands for decision making and opportunities for autonomy among machinists were reduced by the move to numerically controlled equipment.

Opposing these views are those of automation advocates, who emphasize technology's potential for upgrading skill and work requirements. Some suggest, for example, that job complexity and flexibility will increase, requiring more skill and judgment from workers (Levin & Rumberger, 1983; Piore & Sabel, 1984). The high-technology future is seen as enhancing worker autonomy, decreasing machine pacing, and encouraging job enlargement and improvements in general working conditions (Ayres & Miller, 1983).

Both these sets of predictions are incorporated in other forecasts of a bimodal job structure having a small number of highly skilled autonomous workers and a much larger cadre of unskilled machine-controlled workers. In this view, job losses are seen as likely around the middle of the job continuum, with the two ends becoming increasingly polarized (Skinner & Chakraborty, 1982). Support for the bimodal prediction has been found in studies of communications workers (Denny & Fuss, 1983), where the most adverse effects were noted among the least skilled operators and the most positive effects among the highly skilled positions. Majchrzak, Chang, Barfield, Eberts, and Salvendy (1987) reported similar patterns in a study of engineering design jobs.

Important Contingencies

Employment and occupational forecasting is an imprecise art. Available evidence indicates that the effects of programmable manufacturing automation are relatively indeterminate, allowing a wide range of organizational and manpower solutions (Kohler & Schultz-Wilde, 1983; Office of Technology Assessment or OTA, 1984; Warner, 1984). Programmable automation includes a wide range of equipment and systems that can be used in different combinations and in conjunction with conventional equipment. As with other technological innovations, their impacts—both positive and negative—are difficult to foresee in entirety because their application and secondary spin-offs can go beyond that originally envisioned (Bednarzik & Schoepfle, 1985). Simple predictions of the labor impacts of programmable automation should therefore be examined in terms of their underlying assumptions regarding the behavior of firms and individuals involved in the implementation of these technologies, and regarding external conditions such as the state of the economy (Andelin, 1983).

A major factor affecting the actual changes that may occur as more computerized manufacturing equipment is implemented is the rate at which implementation occurs. It is clear that the faster the rate of change, the greater the potential dislocations may be. From this perspective, it appears that earlier visions of massive job losses have been exaggerated. The era of the unmanned factory is still very far away. A 1982 survey of 303 firms in three industries identified by experts as likely users of computerized manufacturing equipment (transportation equipment, electric and electronic equipment, and industrial and metal working machinery) showed limited use of computer-automated equipment (Majchrzak, Nieva, & Newman, 1986; OTA, 1984).

Even firms that have decided to adopt the new technologies often opt for a gradual introduction of programmable automation. This incremental approach is often due to the significant costs involved in acquiring and implementing the new computerized equipment, the lack of data regarding its potential benefits, and the fundamental and extensive factory restructuring that is often needed to accommodate the new technology (Bednarzik & Schoepfle, 1985; Hunt & Hunt, 1983). Such a pattern suggests that changes in employment levels may be dealt with through normal attrition, combined with retraining of the existing work force.

The impact of new technology on workers also depends critically on the human resource philosophies and practices of employers. Different national philosophies have been shown to affect the way in which new technology is implemented. A study comparing the consequences of CNC utilization in British and German manufacturing firms that were matched in terms of size, product, and technology showed the malleable nature of computerized manufacturing technology and the relatively stronger effects of "country" variables (Hartmann, Nicholas, Sorge, & Warner, 1984). Country differences were particularly notable in the larger firms. In general, compared to the British firms, production operator jobs in the German firms had higher skill requirements due to the inclusion of planning and programming activities. Greater differentiation and specialization characterized the job structures in the British firms. Thus, in Britain, but not in Germany, CNC was viewed as having de-skilling, skilling, and routinizing effects.

Obviously, the behavior and decisions of individual employees also affect the employment impacts of technological change. Whether employees seek or avoid retraining, whether they resist or accept coming changes in the condition and nature of employment, whether they are willing to take different types of jobs, and whether they take a reactive or proactive stance toward job-related technology issues are important factors in determining the consequences of new technology.

Finally, employment impacts also depend on alternative employment opportunities available to displaced workers. Overall economic growth is usually the most important factor in determining growth or decline of employment, although the reverse causality is also possible (Bednarzik & Schoepfle, 1985). Thus the Bureau of Labor Statistics suggests that employment growth and technological change will continue to move together over the next decade (Kutscher, 1984). Consequently, job displacement will fall most heavily on regions and industries where alternatives are scarce (Hunt & Hunt, 1983).

The Research Site

The research site for the study described in this chapter was a manufacturing facility located in a large manufacturing and product complex on the west coast of the United States. The manufacturing complex was one of several belonging to a multibillion dollar Fortune 100 corporation involved in the manufacture of electromechanical products.

In the late 1970s, the corporation suffered many serious business and financial problems that posed fundamental challenges to the way the company operated. During this period, the corporation's near monopoly on its major product line was seriously undermined when international competition caused its market share to plummet from almost 100% to about 40%. At the beginning of the 1980s, a new corporate senior management team set its sights on resuming its former position of dominance in the marketplace by cutting costs, streamlining the product development process, and emphasizing product quality. A broad-ranging corporate quality program that involved elements of product positioning, competitive benchmarking, and employee involvement was launched. Each business unit, including our research site, was given the responsibility for planning and implementing its own quality improvement program.

The research site, which we shall call Factory XYZ, produces printed wiring board assemblies (PWBAs) for electromechanical equipment manufactured by the parent corporation and for other external customers. During the time period covered by the study (1985 to 1987), the factory produced over 1 million PWBAs annually. Over 250 types of assemblies, each consisting of several hundred components, were produced by a variety of end products in lots ranging from 10 to 10,000. The majority of these assemblies were designed for specific products by an engineering department in the complex.

At the start of the research project, Factory XYZ had just begun a major effort to move toward becoming a "factory of the future" with the implementation of new computerized manufacturing technologies and the streamlining of its manufacturing process. This initiative was the factory's response to a general corporate challenge to improve its competitive posture in the international marketplace. At that time, manually assembled boards were viewed as excessively labor intensive and error prone. The new technologies were seen as vehicles for reducing the amount of human handling in the assembly process, which in turn would reduce board costs and improve quality. Further, the new technologies comprised a response to the continuing growth in the demand for PWBAs that resulted from the increasing electronic content of most of the corporation's products. Thus the move toward the "factory of the future" was aimed at the multiple goals of cost reduction, quality improvement, and increased production volume.

Methodology and Data Sources

The data for this chapter were drawn from three sources: in-depth qualitative interviews with factory managers, supervisors, and shop-floor personnel; employee rosters obtained from the personnel department and other documents; and surveys of nonsupervisory employees at two points in time: 1985 and 1987. These data were gathered during six site visits to the PWBA factory over the 3-year study period. Each site visit was conducted by a 3-person research team and lasted between 3 and 10 days. The visits were used to construct interviews with key informants to collect archival materials, to administer questionnaires to plant personnel, and to observe the activities of shop floor employees and their supervisors.

Interviews with Factory Managers, Supervisors, and Shop Floor Personnel

Qualitative interviews were conducted with about 40 respondents directly involved in the planning, process, and management of production operations at the plant, including plant management, production personnel, shop floor managers, production and maintenance engineers, and corporate staff.

At each level of the organization, we discussed recent and ongoing technological developments on the shop floor, as well as associated developments in the areas of product design and production planning. Through these interviews, we developed a chronology of recent technological developments and organizational changes that accompanied them. At the management level, we examined the process of evaluation, design, and implementation of the new equipment. Interviews with design and production engineers focused on expectations and reactions to the new equipment, while interviews with shop floor personnel concentrated on job descriptions and characteristics.

Surveys of Nonsupervisory Employees

Questionnaires were administered twice (in March of 1985 and May of 1987) to production personnel on-site at the factory. The first wave of data was collected at the very early stages of technology change in the factory. By the second wave, two years later, much of the initial push toward new technology had been completed.

These questionnaires were used to analyze changes over time in the nature of production-level jobs in the factory, and the reactions of workers to various aspects of their work environment. The survey questionnaires were developed through a combination of existing instruments, job observations, and personal interviews. From standardized surveys of employees—including the Organizational Assessment Instrument (Van de Ven & Ferry, 1980), the Job Diagnostic Survey (Hackman & Oldham, 1974), and the Michigan Survey of Organizations (Taylor & Bowers, 1972)—we selected questions relating to job requirements and characteristics, interdependence among employees and work groups, job satisfaction, feedback, and training. Additional questions enumerating the specific tasks of production personnel were then created, based on interviews with production floor employees and their supervisors and supplemented by informal job observations of employees at different stations throughout the factory. Finally, new questions were developed to assess participation in the technology implementation process and reactions to the new equipment implemented in the plant.

The survey population consisted of all nonsupervisory production floor employees, as listed on personnel rosters provided by the company. A total of 416 surveys were distributed in 1985, with approximately 15% in a Spanish version. The response rate for the 1985 administration was 46%, a total of 192 returns. A slightly modified version of the questionnaire was administered in 1987. The population for the second administration was somewhat smaller than that of 1985 (n = 272), as the factory had reduced payroll levels through early retirements, attrition, and the expiration of some temporary positions. The return rate for the second administration was much higher (78%, n = 213) than the earlier administration. This improvement was attributed to greater familiarity with both the survey and the researchers, coupled with better follow-up with supervisors during the administration period.

Employee Rosters and Other Plant Documents

A third data source for this study consisted of the 1985 and 1987 employee rosters from the personnel department. Rosters were provided for all nonsupervisory personnel at the plant in both years. The rosters listed work-area supervisor assignments and job titles for all personnel assigned to the production floor. Production personnel included all hourly production workers, along with salaried employees who staffed the maintenance of test areas. The employee rosters from 1985 and 1987 were compared to examine the nature of population changes in the factory and to identify shifts in work group assignments and staffing structures.

The Assembly Process and Factory Work Groups

Factory XYZ's plan to improve the quality and efficiency of its operations centered on streamlining its production process by implementing new technologies in manufacturing PWBAs. Two major types of assembly processes took place in the factory: "conventional" and "automatic." Since the factory had started using some automated insertion equipment in the 1960s this distinction between conventional and automatic processes was not absolute, but rather a matter of degree. The conventional assembly processes included both purely manual operations and subprocesses that used islands of computer-controlled equipment. The automatic process was conducted almost completely by a linked series of computerized equipment, although a very limited set of operations was still conducted manually. In 1985, about 20% of the boards were assembled manually, about 40% were assembled using the semiautomatic islands, and about 40% were assembled using the automatic process. Plans for the factory of the future saw production moving much more heavily toward the fully automatic processes, which were envisioned as becoming increasingly flexible and very closely linked to the board design processes.

PWBA assembly begins with the receipt of "bare boards" from another of the corporation's factories and the preparation for assembly of numerous types of externally purchased electronic components (e.g., axials, heat sinks, transistors). The basic objective of the manufacturing operations is to insert a variety of electronic components into predesignated places on a bare board. Insertion activities included a variety of processes and technologies—manual operations using such tools as wire cutters, pliers, soldering irons; manual operations conducted with machine assistance; and operations done automatically by a series of computer-controlled machines. Electrical connections between the components were made manually or by a wave soldering process. The PWBAs were inspected at various subassembly stages in the production process and were tested electronically at the end of the assembly process. After passing final inspection, the assemblies were packaged and shipped to the various customers. Figure 3.1 illustrates the major steps involved in the conventional and automatic processes in 1985.

Conventional Assembly

At Factory XYZ, PWBAs assembled by the conventional process moved through a number of work groups corresponding to the functions represented in Figure 3.1. Some of these work groups (i.e., Intake, Offline, and Post Solder Assembly) were mostly involved in manual operations. Intake workers were responsible for the receipt of all board materials. They stamped identifying information on the bare boards and prepared components for assembly using traditional tools such as pliers, wire cutters, and soldering machines. These areas handled operations that were not designed for automatic processes, either because the board design preceded the installation of automated machines at the factory or because characteristics of certain components on the boards (e.g., size or heat sensitivity) required manual handling by skilled assemblers.

The Auto-Insert (A/I) group in the conventional assembly process involved the coordinated efforts of workers and islands of automated equipment. Workers operated equipment that had been programmed specifically for each assembly. The A/I equipment consisted of three sets of machinery. The first set, called the sequencers, put component parts in the required sequence and tested each part following programmed instructions. The second set, called the DIP insertion machines, installed integrated circuits called dual-in-line packages (DIPs) into the bare boards. The third set, called the variable center distance machines or VCDs, inserted the sequenced component parts into the boards. The A/I workers, therefore, were essentially tenders or operators of stand-alone programmed machinery.

In contrast to both the manual and Auto-Insert work groups, the Progressive Assembly Line (PAL) resembled the traditional model of a manufacturing assembly line. In the PAL area, workers sat in front of a moving conveyor line, with a supply of component parts arrayed in trays around them. Their task was to insert the appropriate parts manually into the boards, according to diagrammed instructions that they kept at their work stations.

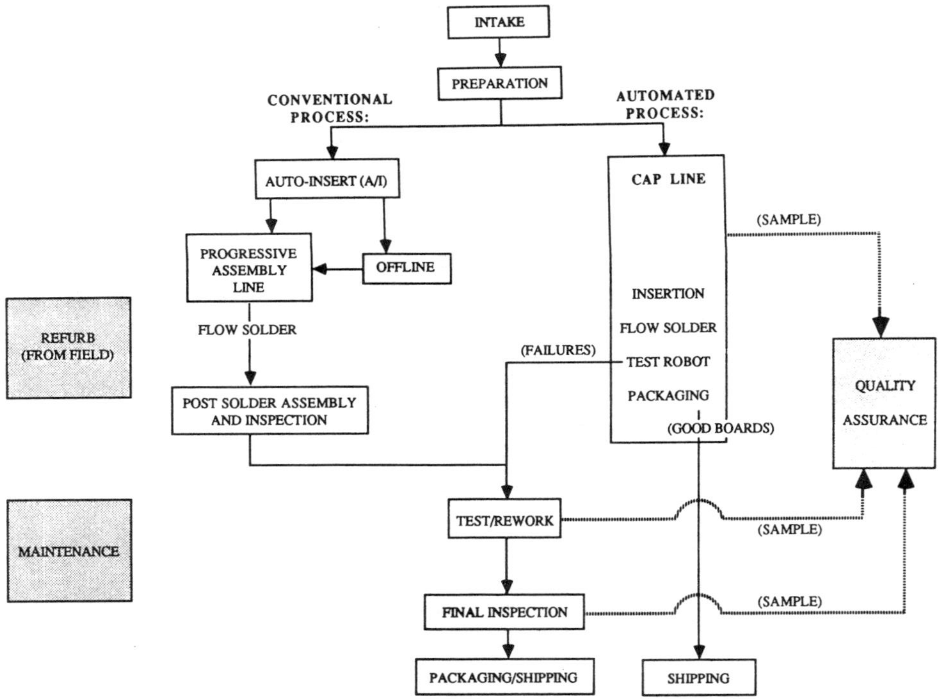

Figure 3.1. PWBA Processes and Major Work Groups – 1985

NOTE: Unshaded boxes indicate the PWBA assembly process/work groups; shaded boxes indicate production support processes/workgroups that were not directly part of the assembly process. Solid lines indicate that all outputs moved on to the next process step; broken lines indicate that only a sample moved to the next step shown.

Board quality in the conventional process was maintained by visual inspection and quality assurance activities conducted at various points in the assembly process. In 1985, inspection was conducted by a separate work group whose members were stationed at each step in the process. Inspectors examined the boards for visible defects, such as bent leads, misoriented components and missing parts.

The test area at the end of the line involved a series of computerized electrical tests that checked the functionality of individual components as well as that of entire boards. While all the testing equipment was computerized, it varied widely in terms of complexity and diagnostic capacity. The tests performed varied across PWBAs. The test area workers were divided into two subgroups that differed in job difficulty—the hourly technicians conducted the simpler board tests while the salaried testers performed the more complex testing procedures. Most conventional assemblies underwent an in-circuit test, which verified that the components were not defective and had been placed properly, and functional tests, which tested how the overall assembly performed. About half of the assemblies received system tests, assessing their performance in the actual products for which the PWBAs were intended, or emulator tests, using machines that simulated the actual system. The type of testing conducted was related to the life cycle of the board. Early in their development, all boards were tested extensively, undergoing examinations that included functional and emulator tests. As the board matured and the design stabilized, fewer tests were needed.

At the end of the entire process, assemblies were sent to inspection, where they were visually inspected for general workmanship configuration and completion of solder joints. Inspection activities were also supplemented by a separate Quality Assurance unit (see production support groups, described below).

Automatic Assembly

The automatic process was conducted on a computer-controlled Continuous Assembly Process (CAP) line that had been installed in the factory in 1984, just before the study started. The CAP line was a microcosm of the conventional production process, and its crew handled all operations performed by separate work groups in the conventional process. The CAP line was essentially a paced conveyor line along which a series of computer-controlled machines were stationed. The boards' route through the CAP area took them through the following equipment (listed in assembly line order):

- two automatic DIP (dual-in-line packages) insert machines,
- axial insertion machines,
- bench assembly for the insertion of parts that could not be automatically inserted,
- flow solder and wash stations,
- a test robot that performed a functional test and sorted good boards from bad, and
- a packaging station that inserted good boards into shipping containers and stacked them on carts for transfer to the shipping area.

As previously noted, the designation of conventional and automatic assembly was not an absolute dichotomy, but rather a matter of degree. Although over 90% of the operations conducted on the CAP boards were automatic, a bench assembly area did exist to handle parts that could not be automatically inserted. Some boards, for example, were designed with components on both sides, and the CAP line could only handle insertion on one side. The automatic test function conducted by the test robot on the CAP line was also supplemented by additional functional diagnostic tests for the small percentage of boards that failed.

Production Support

In addition to the main production work groups, there were three major supporting work groups on the factory floor: the Refurbishment (Refurb) unit, the Quality Assurance unit, and the Maintenance group. The Refurbishment group was composed of highly skilled technicians responsible for repairing or upgrading boards that were returned from the field, using largely manual and craftlike procedures. The Quality Assurance group was responsible for the inspection and evaluation of a sample of boards and subassemblies at various points in the production process. The other important production support group was Maintenance, which was divided into two subgroups. The Test Maintenance

Table 3.1
Major Work Groups and Functions

Work Group	*Function*
Intake and Preparation	Receipt of bare boards and engineering documentation; inspection of incoming components and preparation of parts for assembly.
Progressive Assembly Line (PAL)	Manual insertion of drop-in parts into boards moving on a conveyor; operation of flow solder machines to set electrical connections on boards.
Automatic Insertion (A/I)	Setup and operation of A/I equipment: sequencers, DIP insertion machines, and VCDs.
Offline	Manual assembly operations that could not be done by automatic machines or by people working on the PAL conveyer.
Post Solder Assembly	Hand-soldering of parts that could not be accomplished by the flow solder; manual reworking of defective boards.
CAP Line	Operation of the computerized CAP line, which involved loading bare boards at the start of the line, maintaining adequacy of supplies of component parts at each insert station, and board inspection as the boards progress down the line.
Testing (Hourly & Salaried)	Electronic testing of in-circuit components and assembly functionality, using a variety of computerized tests.
Refurbishment	Repair and upgrading of boards returned to the factory from customers.
Inspection and Quality Assurance	Visual inspection of subassemblies and finished PWBAs.
Packing and Shipping	Packing of finished assemblies.
Maintenance	Test equipment maintenance and production equipment maintenance.

group was responsible for the electronic test equipment, while the Production Maintenance group was responsible for all other equipment on the production floor, including conventional and automated electromechanical equipment.

Changes Toward the Factory of the Future

The factory's push toward becoming a "factory of the future" involved a number of simultaneous innovations. Most visible, but not necessarily most critical, was the increasing automation and computerization of technology on the factory floor. As noted previously, Factory XYZ already had some limited applications of programmable technology in place, and this push was seen as a logical extension of small initiatives that had begun earlier. However, the implementation of new manufacturing technologies was only one of several fundamental changes that were part of the corporation's overall strategy for increased competitiveness in the global marketplace. Broad-ranging shifts in corporate culture and philosophy that placed a renewed emphasis on quality throughout all facilities were also initiated by top corporate management. Factory XYZ participated in these shifts at the same time that the technological initiatives were being implemented. Finally, the competitiveness strategy involved reorganizing and rationalizing multiple functions within the plant as well as among various corporate facilities. In Factory XYZ, a variety of basic organizational changes to the plant's work group structures were implemented over the same period.

New Manufacturing Technologies

As described briefly above, Factory XYZ had begun a new initiative of converting to more automated and computer-controlled processes in the early 1980s. During the three-year period covered by this study, several new technologies were implemented toward the joint objectives of cutting costs, improving quality, and increasing production capacity.

The investment in new technologies totaled about $12 million during these three years—a small proportion of the factory's production volume, but nevertheless a significant increase over previous levels of investment.

The most notable technologies implemented were the conversion of assembly instructions from paper tape to computerized forms and the introduction of a second CAP line with robots. In addition, other experimental technologies, both computerized and conventional, were introduced into the plant during this period. These new technologies basically represented a new stage in the factory's efforts to mechanize functions that had traditionally been accomplished manually. None of them involved radical changes in the basic nature of the PWBA assembly process.

In general, the process of introducing these new technologies into the plant was accomplished relatively smoothly. Although about 75% of the plant employees were represented by the Amalgamated Clothing and Textile Workers Union (AFL-CIO), relationships between management and union had historically been friendly. Plant management worked closely with the union in paving the way toward acceptance of the new automation. Communication meetings, introduced into the factory culture in the early 1980s, provided a natural forum for giving early notification to the workers about the impending changes.

Thus, by the time the study started, the threat of international competition was well understood at the plant, and plant workers were convinced that automation was the way to protect the majority of the jobs in the factory. The stage was set for implementing various productivity measures, including new forms of technology. The first survey revealed that the employees were optimistic about the new technology's benefits for the company as a whole and for their own jobs as well (Nieva, Gaertner, Argoff, & Newman, 1986).

Conversion of Assembly Instructions to Computerized Form

The process of creating software programs to run the new automated equipment was a new form of interaction between the manufacturing and engineering divisions at the site. Traditionally, hard-copy assembly instructions for each PWBA type were developed in a documentation section within the factory. Using a variety of inputs, such as the PWBA's bill of materials, the engineering drawing of the board design, and other auxiliary artwork, the documentation specialists created instructions that specified: (1) the board's setup sequence, (2) how to prepare parts and distribute them to work stations, (3) instructions for assembly, (4) machine setup, and (5) change and rework procedures. These documents were used by the factory's production staff to guide activities at the workstations. In addition, the documentation staff used them to develop paper tape programs that were physically inserted into the Auto-Insert machines and CAP line.

At the start of the study, the process of changing over from paper tape programs to computer software had just begun. The first set of computer programs for the Auto-Insert machines had just been completed by the software engineers. Work had just started on the CAP software and on developing a component data base that contained all the basic attributes of each component used in the assemblies. The software conversion effort was largely completed by the end of the study.

Introduction of a Second CAP Line, with Insertion and Test Robots

The first CAP line was introduced into the factory in 1984, just before the study began. The CAP line was implemented as a way to improve board quality by reducing errors attributable to human handling (e.g., misplaced components, missing parts, and misoriented parts). In addition, the CAP was expected to reduce the number of workers responsible for assembly by making one CAP worker responsible for overseeing the operation of four or five machines on the line. In 1986, a second CAP line was introduced that was longer and involved more complex technologies than the first. In addition to two Auto-Insert machines, it had a surface-mounted device (SMD) machine, a pick-and-place robot, and a test robot.

Miscellaneous Process and Equipment Improvements

In addition to the major types of new technologies implemented in the factory, the push toward greater efficiency prompted a variety of improvements in the assembly process. The factory experimented with various pieces of automated equipment that did not remain on the floor for very long. Solstice was a computerized system that was designed to track individual boards as they were routed through the production process. It was first introduced in the testing area in late 1984, but various problems caused the system to be removed from the factory in 1986. Other experiments included a vision system for the CAP line intended to recognize board type automatically, and "single station builds," small islands of automation that allowed several insertion processes to be performed by one worker. Finally, pieces of conventional equipment, (e.g., additional conveyors to the flow solder stations) were implemented without changing the nature or flow of the basic assembly process.

Changes in Culture and Manufacturing Philosophy

A multilevel Quality Improvement Program was an important part of the corporation's push to regain its lost position of dominance in the marketplace. A central theme of this program was the message that each individual had a responsibility for product quality. At the factory, the program was operationalized in a variety of ways. All key managers participated in Quality Improvement workshops. Regular employee communications meetings focused on quality and were institutionalized at several hierarchical levels. In many work groups, especially the CAP, supervisors began to implement the concept in terms of team problem-solving and worker involvement in day-to-day operations.

Another key element of the quality improvement push at the factory was the idea of building quality early in the process. Quality was redefined as an ongoing process, rather than a status to be verified at the end of the line. In specific terms, this meant that inspection and quality assurance activities were shifted from the tail end of the manufacturing process and incorporated into all earlier parts of the production cycle. Traditionally, inspection and quality control functions had been conducted by groups separate from production, and mostly only toward the end of the production process. With the new approach, inspection became part of the job definition of all production personnel, rather than the purview of a unit separate from the production process.

Shop Floor Organizational Changes

Important changes in the organizational structure were also implemented as part of the company's efforts to increase competitiveness during this period. In general, cost consciousness led to efforts to consolidate redundant functions across several plants. In addition, efforts were directed at reducing the level of support activities within factories—i.e., activities not directly involved in production. At Factory XYZ, the following changes were made to streamline the structures and responsibilities of several work groups:

1. *Intake and Preparation Group.* As part of the general effort to consolidate across factories, part of the Intake and Preparation group was centralized in 1986. The Receiving subgroup, in charge of order processing and inventory control, was moved to a central receiving department outside of Factory XYZ. The Intake and Preparation group retained the inspection and preparation functions for incoming components.
2. *Refurbishment and Offline Groups.* Two groups involved in manual assembly were disbanded. The Refurbishment unit, which was responsible for repairing defective boards that failed tests at the end of production or that were returned by clients, was dissolved, and the function was moved to Canada. This change was part of the overall plant strategy to reduce costs and emphasize quality. Since the Refurbishment function depended on highly skilled technicians who could diagnose the sources of a board's malfunction, management decided to move this labor-intensive function to Canada where labor costs were more favorable. Moving the refurbishment function away from the factory also reinforced the factory's shift toward a production philosophy that emphasized in-process inspection and testing and away from error-checking at the end of the production process.

   The Offline group was also dissolved. This group had been responsible for mounting large and unwieldy components, such as heat sinks and transformers, that were not well suited to the traditional insertion machines on the production floor. By 1986, many of these parts could be inserted automatically by the pick-and-place robot on the second CAP line. Insertion of the remaining parts that could not be handled automatically was assigned to Post Solder Assembly, the remaining manual assembly group.
3. *Inspection.* The final organizational change on the production floor was the dissolution of the Inspection group. It was originally assigned a watchdog role that was separate from the manufacturing process, but later the inspection function was incorporated into the other production (PAL, A/I, and CAP) and test groups. This change was in keeping with the new production philosophy that viewed inspection and quality control as part of the production process rather than as a separate activity.

Summary of Factory Changes

The period covered by the study, 1985 to 1987, was a turbulent one for Factory XYZ as well as for the entire corporation. Multiple types of change, often related, were being implemented in different parts of the factory. Table 3.2 presents a graphic summary of the work groups affected by each of the major types of changes discussed above. Acting in concert, these changes had major impacts on the factory's employment levels and the characteristic of the jobs on the factory floor (these impacts are discussed in the following section). From the multiplicity of simultaneous forces operating on the factory, it is clear that no simple explanations for the employment and job impacts observed will be readily available. The discussion of these impacts, therefore, consists of an interpretive synthesis of the information gathered on the various interconnected forces at work during this period.

Changes in the Number of Jobs, Job Structure, and Job Quality

In this section, results are presented regarding the changes found at the plant between 1985 and 1987 in the three variable sets hypothesized to be vulnerable to the types of organizational and technological changes experienced at Factory XYZ; the number of jobs, job task structure, and job quality. For each of these topics, the specific quantitative data sources used in the analyses are discussed prior to the presentation of results. In each case, discussion of quantitative results is interwoven with information gathered through the qualitative interviews.

Number of Jobs

At base, the changes occurring at Factory XYZ were designed to improve the company's international competitive position and reduce costs. Similar efforts to improve company competitiveness have led many scholars (e.g., Ayres & Miller, 1982) and employee advocates to

Table 3.2
Summary of Factory Changes and Groups Affected

Types of Change	*Intake*	*PAL*	*A/I*	*Offline*	*Post Solder*	*CAP*	*Test*	*Refurb*	*Inspec-tion*	*Main-tenance*
Technology										
Assembly Software			X			X				X
2nd CAP Line				X		X				X
Miscellaneous						X	X			
Culture & Philosophy	X	X	X	X	X	X	X	X	X	X
Shopfloor Organizational Changes	X			X				X	X	

question whether such improvements for the company are achieved at the expense of their workers. This study examined the impact of the productivity efforts at Factory XYZ on employment levels among its production staff.

Analyses of changes in factory employment levels were based on company personnel rosters obtained in 1985 and 1987. The rosters contained the names, work area assignments, and supervisor assignments for all active personnel on the production floor. In addition, the lists included information regarding employment-related actions, such as temporary layoffs, recalls, retirements, and terminations.

Results

As expected, the efforts toward increasing production efficiency entailed substantial reduction in the number of job slots in the factory. From 1985 to 1987, a period in which production levels remained relatively flat, the number of plant production employees dropped from 412 to 296. Included in this count were nonsupervisory hourly and salaried employees on the production floor, Test personnel, and Maintenance staff. This represented a loss of 116 positions, or 28% of the original number of jobs, over the two-year period studied.

It should be noted that the period during which the study was conducted was a very volatile one, not only for the corporation and Factory XYZ, but for the electronics industry in general. The factory's dip in employment from the 1985 levels largely mirrored the experience of the electronics industry. For example, the value of shipments by manufacturers of electronic components and related lines of business grew from almost $10 billion in 1975 to a peak of over $45 billion in 1984, but the curve started to reverse itself in 1985. The high cost of producing electronic components, combined with intense foreign competition from countries with low labor costs, resulted in a recession in U.S. operations throughout the industry (U.S. Department of Commerce, 1988). Employment levels in the industry as a whole followed the production curve.

Although there was a large decrease in the number of production jobs at Factory XYZ, the reduction of 116 job slots was accomplished with very limited adverse impact on individual employees. Only 9 of the 116 displaced employees were involuntarily terminated. Reductions were accomplished primarily through transfers to other related plants within the complex and through retirements. In addition, a few temporary

Table 3.3
Status of Employees Displaced from XYZ Jobs

Status	*Number*
Transfer to other plants within the complex	33
Retirement/voluntary termination	26
Temporary/assignment ended	9
Leaves of Absence/disability	14
Terminations	9
Unknown*	25
	116

*These were salaried workers whose status after leaving their jobs at the factory could not be tracked. Union records and rules for job posting and seniority decisions required structured tracking of hourly employees. Salaried workers, however, moved more freely between jobs in the various divisions, and personnel records for these employees were not easily available.

positions were dissolved, and no substitutes were put in their place. In sum, this reduction in force was conducted with a strong concern for minimizing the trauma to the incumbent work force. These results are in keeping with Hunt and Hunt's prediction (1983) that most job loss in automating plants can be handled gradually through normal attrition, transfers, and retirements. Table 3.3 shows the eventual outcomes for the employees who occupied the positions that were closed out.

The reductions in size varied widely across work groups. Table 3.4 shows the number of employees in each work group in 1985 and 1987, presented in an order that will facilitate the discussion. In general, the greatest number of job losses were found among the manual assembly groups (Post Solder Assembly, Offline, and Refurbishment), which dropped from 89 jobs in 1985 to 26 jobs in 1987. (It should be noted that the bulk of this decrease can be accounted for by the decisions to disband the Offline and Refurbishment functions within the factory.) In contrast, the number of assembly jobs involved in some form of automation (Auto-Insert, PAL, and CAP) increased over this period from 108 to 131, an increase of about 20%.

The other groups that suffered the largest losses were support groups not directly involved in assembly (Intake, Inspection, and Test). This pattern was consistent with the company's general efforts at streamlining its processes and cutting "nonproductive" costs. Proportionately larger losses were found among the hourly testers, compared to their salaried counterparts.

Table 3.4
Number of Job Slots by Work Group in 1985 and 1987

Work Group	*1985*	*1987*	*Difference*
Intake			
Receiving	28	–	–28
Preparation and Inspection	23	14	–9
Assembly Groups Using Automated Technology			
PAL	22	31	+9
Day Shift Auto-Insert	35	28	–7
Day Shift CAP	18	35	+17
Night Shift CAP/Auto-Insert	33	37	+4
Manual Assembly Groups			
Post Solder Assembly	52	26	–26
Offline	17	–	–17
Refurbishment	20	–	–20
Inspection and Quality Assurance	22	–	–22
Test			
Hourly	34	19	–15
Salaried	61	51	–10
Maintenance	20	18	–2
Packing and Shipping	27	37	+10
Total	412	296	–116

The reduction among the manual assembly jobs, either by dissolving a unit or reducing its size, and the increase in the automated assembly jobs were clearly attributable to the ascendance of automated technologies in the production process. Technology appeared to be related to job losses in one other category: hourly Test technicians. Over this period, the bulk of the testing activity moved away from the simple tests conducted by the hourly technicians toward increasingly sophisticated test equipment operated by higher-level, salaried Test workers. In 1987, the number of hourly job slots in the Test group had been decreased by nearly half.

Other factors are needed to explain the job reductions observed in the other support work groups, factors that appear to be specific to each work group. For the Intake group, reductions were made possible by improvements in relationships between the factory and its component vendors. The factory's efforts to work more closely with its vendors over this period resulted in higher quality components, reducing the effort required for inspection and preparation. These closer vendor relationships also resulted from components delivered in packaging that was designed to the factory's specifications, again minimizing the handling necessary prior to assembly.

The dissolution of a separate Inspection and Quality Assurance unit and the substantial reduction among the Test staff were attributable, in large part, to the shift in culture and manufacturing philosophy that occurred throughout the entire corporation. As noted previously, a renewed corporate emphasis on quality improvement led to the integration of responsibility for quality into all steps of the production process. This implied that a separate unit responsible for quality assurance was no longer desirable. Further, earlier error detection implied lower levels of effort required of the Test area at the end of the line. The reduction in Test staff was facilitated by the increased diagnostic capacities of the computerized testing equipment. The manufacturing philosophy emphasizing quality early in the production process and the increased technological sophistication of the testing equipment combined to help the factory to achieve its joint goals of cutting costs and improving quality.

Job Task Structure

In addition to issues of job displacement, another focus of this study was to examine more subtle changes in job duties and skill requirements resulting from the factory's improvement efforts and the implementation of new technology. Did the jobs at the factory change in major ways? Did jobs become increasingly specialized and fragmented, or was there evidence of job broadening or even "job blending"?

In this study, changes in job task structure were examined through employee reports of time allocation to tasks. In the 1985 and 1987 surveys, nonsupervisory employees were asked to indicate the number of hours that they typically spent on each of the 18 job tasks shown in

Table 3.5
Job Tasks in Employee Survey

1.	Loading or unloading PWBAs or feeding components
2.	Preparing boards or components for assembly
3.	Getting parts to correct locations on the production floor
4.	Hand-inserting components into a *moving* board (with or without tools)
5.	Hand-inserting components into a *stationary* board (with or without tools)
6.	Running the flow solder machine
7.	Inspecting the boards or components
8.	Monitoring the operation of the machine
9.	Setting, adjusting, or correcting production equipment
10.	Repairing equipment breakdowns
11.	Conducting preventive maintenance
12.	Debugging programs when equipment malfunctions
13.	Testing boards or components
14.	Troubleshooting or diagnosing PWBAs
15.	Repairing/reworking boards or components
16.	Making special tools/equipment or parts
17.	Training other employees to operate equipment
18.	Reporting the status of products or the production process

Table 3.5. This task list was developed after a series of job observations and interviews with factory staff and supervisors, and it contains all the major tasks performed on the production floor. Since the individual reports of hours worked added up to different totals, the hours reported by each employee were converted to percentages of total time that each employee spent on specific tasks.

The mean percentages of time spent by each work group on each task in 1985 and 1987 were compared to determine the relative emphasis on the various assembly tasks at the start and end of the study. Since jobs were largely defined in terms of work group membership, these shifts in relative emphases were interpreted as changes in the nature of the jobs themselves.

For each work group, two indicators of change were calculated. As a measure of overall *magnitude of job change* between 1985 and 1987, Cronbach and Gleser's (1953) *D* was calculated. *D* is basically a measure of profile dissimilarity that simultaneously takes into consideration changes in level, dispersion, and slope between the measures in the two years. The larger the *D*, the greater the change in the way work groups allocated time to tasks.

The second indicator, *work group task variety*, was calculated as the number of "primary" tasks on which at least 10% of each work group's time was spent. The 10% figure was judged to be a substantial portion of the group's work time. The size of this indicator represented the diversity of production activities performed by the group as a whole. The number of primary tasks in 1985 was compared to that of 1987.

To supplement this measure of task variety at the group level, parallel indicators of *individual task variety* were calculated. For each respondent, we counted the number of primary tasks, again defined as tasks on which the individual spent at least 10% of his or her work time. The average individual number of primary tasks per work group for 1985 and 1987 was then obtained, and differences were examined to determine whether the group-level changes were paralleled at the individual level.

Results

The analyses compared the extent of job change experienced by the different work groups. Where substantial changes were found, the nature of these changes (i.e., shifts in time allocated to specific tasks) was explored. Finally, the discussion compares the changes in task variety observed at the group and individual levels of analysis.

Table 3.6 shows the 1985 and 1987 average percentages of the work group's time spent on each task, and changes in time allocation to tasks. Figures of 10% or larger, referred to in the discussion as substantial proportions of time, are underlined. The table also shows *D* measures for each group, indicating the overall *magnitude of job change* between 1985 and 1987. Finally, for both years, it shows measures of group-level *task variety*.

Because of the small numbers of respondents, employees in the three manual assembly groups (Offline, Post Solder, and Refurbishment) were combined into one group, called Benchwork. The tasks in these three work groups were judged to be sufficiently similar to make this grouping reasonable. In addition, the data for hourly and salaried testers are presented separately because of the large differences obtained for these subgroups.

Extent of Change

The largest changes in terms of work-group distribution of time occurred in the Intake group (D = .53) and hourly Test (D = .47).

Table 3.6
Percent of Work Group Time Allocated to Tasks in 1985 and 1987

Job Tasks		*Intake* 85	*Intake* 87	*Intake* Diff	*PAL* 85	*PAL* 87	*PAL* Diff	*Auto-Insert* 85	*Auto-Insert* 87	*Auto-Insert* Diff	*CAP* 85	*CAP* 87	*CAP* Diff
	N =	33	22	−11	6	23	17	17	16	− 1	17	45	28
1. Load/unload PWBAs		2	0	− 2	14	9	− 5	6	13	7	18	19	1
2. Prepare for assembly		20	64	44	2	12	10	7	11	4	8	7	− 1
3. Get parts to location		8	4	− 4	24	6	−18	2	1	− 1	8	5	− 3
4. Insert to moving boards		0	0	0	18	33	15	1	5	4	18	12	− 6
5. Insert to stationary boards		0	0	0	17	26	9	1	5	4	5	5	0
6. Operate flow solder machine		0	0	0	0	0	0	1	2	1	8	4	− 4
7. Inspect boards/components		24	6	−18	17	9	− 8	20	14	− 6	13	20	7
8. Monitor a machine		16	0	−16	0	0	0	33	20	−13	9	11	2
9. Adjust production equipment		3	4	1	6	2	− 4	11	8	− 3	2	4	2
10. Repair equipment breakdowns		0	9	9	0	1	1	3	2	− 1	1	0	− 1
11. Preventive maintenance		0	0	0	0	0	0	2	0	− 2	1	2	1
12. Debug programs		1	0	− 1	0	0	0	0	3	3	1	0	− 1
13. Test boards/components		12	6	− 6	0	0	0	2	0	− 2	0	0	0
14. Debug PWBAs		0	0	0	0	0	0	2	0	− 2	0	0	0
15. Repair/rework boards		0	7	7	0	3	3	4	11	7	5	8	3
16. Make special tools/parts		0	0	0	0	0	0	0	0	0	0	0	0
17. Train others to operate equip.		6	0	− 6	0	1	1	1	1	0	1	1	0
18. Report status of production		7	0	− 7	3	0	− 3	4	2	− 2	2	2	0
Magnitude of change:	*D* =		0.53			0.29			0.20			0.11	
Task variety – group level	1985		4			5			3			3	
(No. of tasks with 10% or more of time allocated)	1987		1			3			5			4	
	Diff		−3			−2			+2			+1	

(Continued)

Table 3.6 (continued)

Job Tasks	Bench-work 85	87	Diff	Test (hourly) 85	87	Diff	Test (salaried) 85	87	Diff	Maintenance 85	87	Diff	Total 85	87	Diff
N =	34	26	− 8	25	13	−12	30	44	14	15	14	−1	192	218	26
1. Load/unload PWBAs	4	12	8	2	7	5	1	3	2	0	1	1	4	9	5
2. Prepare for assembly	11	13	2	0	1	1	1	1	0	0	0	0	6	12	6
3. Get parts to location	1	3	2	0	0	0	1	0	− 1	0	1	1	3	3	0
4. Insert to moving boards	4	3	− 1	0	5	5	0	0	0	0	0	0	3	8	5
5. Insert to stationary boards	30	15	−15	0	0	0	0	0	0	0	0	0	6	7	1
6. Operate flow solder machine	4	5	1	0	5	5	0	0	0	0	0	0	1	2	1
7. Inspect boards/components	13	12	− 1	2	22	20	5	4	− 1	0	1	1	14	12	− 2
8. Monitor a machine	1	2	1	6	0	− 6	6	3	− 3	5	12	7	9	6	− 3
9. Adjust production equipment	1	3	2	1	1	0	2	1	− 1	11	16	5	4	4	0
10. Repair equipment breakdowns	1	0	− 1	1	0	− 1	1	1	0	42	32	−10	5	4	− 1
11. Preventive maintenance	4	0	− 4	1	0	− 1	0	1	1	16	19	3	3	2	− 1
12. Debug programs	1	1	0	1	0	− 1	0	1	1	5	8	3	1	1	0
13. Test boards/components	0	0	0	33	7	−26	39	58	19	2	1	− 1	15	14	− 1
14. Debug PWBAs	0	0	0	27	2	−25	31	22	− 9	3	1	− 2	10	5	− 5
15. Repair/rework boards	18	31	13	24	44	20	4	2	− 2	1	0	− 1	9	10	1
16. Make special tools/parts	6	0	− 6	0	0	0	0	0	0	5	3	− 2	1	0	− 1
17. Train others to operate equip.	0	0	0	1	0	− 1	5	1	− 4	7	4	− 3	3	1	− 2
18. Report status of production	1	0	− 1	2	6	4	3	1	− 2	4	1	− 3	3	1	− 2
Magnitude of change: *D* =		0.23			0.47			0.22			0.15			0.12	
Task variety—group level (No. of tasks with 10% or more of time allocated) 1985		4			3			2			3			3	
1987		5			2			2			4			4	
Diff		+1			−1			0			+1			+1	

NOTE: Percentages greater than 10% are underlined for ease of interpretation.

Substantial changes were also found in PAL (D = .29), Benchwork (D = .23), salaried Test (D = .22), and Auto-Insert (D = .20). The smallest changes occurred in the CAP (D = .11) and Maintenance (D = .15) groups.

The stability in the latter two groups is easily understandable since these work areas basically maintained the same sets of responsibilities over this time frame. The CAP group was already working on the most highly automated piece of equipment on the floor, and the addition of one more CAP line did not change the basic nature of the job. Similarly, the responsibilities of the Maintenance group remained essentially the same, although there was an increase in the quantity of equipment on the floor.

Nature of Change

Where substantial changes were observed, the character of the changes appeared to vary widely across work groups. In some work groups, the distribution of the group's time suggests a narrowing of focus over time, while the opposite pattern was found in other groups. For some groups, the changes suggest a job redefinition rather than a shift in breadth. It is necessary to examine the nature of these changes in order to make reasonable attributions to factors responsible for the change. For example, in the Intake and Benchwork groups, changes in work group time allocation can be attributed largely to the organizational changes described earlier, changes that removed substantial portions of their traditional responsibilities. Further, since Benchwork comprised the most traditional manual assembly activities, the introduction of increasingly computerized equipment was expected to affect the nature of these jobs. Where no such major causal factors were involved, the explanations became less straightforward, involving a mix of technology and manufacturing practices. For several groups, the changes seem attributable to increasing routinization of the new technology rather than to fundamental alterations in job responsibilities. More specific discussion of these changes follows.

Job Narrowing—Intake and PAL

The changes in both the Intake and PAL groups took the basic character of much more narrowly focusing the duties of the groups. In 1985, the Intake group spent only 20% of its time on its core function, preparation for assembly. During this early period, the Intake group had four "primary" tasks (preparation for assembly, inspection of boards/components, monitoring a machine, and testing boards/components). In 1987, the majority of time (64%) was spent on its core task, preparation for assembly. In large part, this shift reflected the reorganization that moved the inspection function for received components to a centralized location outside the plant.

The PAL group's change in time allocation also showed a basic narrowing of focus and a reduction of time spent on auxiliary activities. In 1985, only about a third of the time (35%) was spent on the group's main function, insertion of parts to boards. Substantial amounts of time had been spent on five tasks (load/unload PWBAs, get parts to location, insert to moving and stationary boards, and inspection). By 1987, the time spent on loading, getting parts to the appropriate location, and inspection had dropped off. The reduction in worker time spent on the physical moving of boards and parts was in line with general efforts in the factory to make the materials handling function more efficient. These efforts consisted largely of procedural changes and increased use of traditional conveyor technology. The PAL group's time was then concentrated on preparation for assembly and actual insertion to moving and stationary boards.

Job Broadening—Auto-Insert

While the Intake and PAL jobs basically became narrower in focus, the reverse pattern was observed for the Auto-Insert group. In 1985, its time was concentrated on three tasks: inspecting boards and components, monitoring a machine, and adjusting production equipment. By 1987, Auto-Insert workers were spending substantial amounts of time on these tasks, as well as on others on which very limited amounts of time were previously spent (i.e., loading and unloading PWBAs, preparing for assembly, and repairing boards). Increase in the repair and rework task is particularly interesting, as it appears to reflect the new philosophy of integrating these functions into each work stage and moving them as early in the assembly process as possible.

Job Redefinition—Benchwork and Test

Complex patterns of task changes were seen for the Benchwork and Test groups. For these groups, the changes suggest that the jobs were basically redefined. Benchwork members concentrated on substantially the same tasks (preparing for assembly, inserting to stationary boards, inspection and repair of boards), but large shifts in time distribution were observed for some of these tasks. Benchworkers reduced the time spent on inserting parts to stationary boards, and increased the time spent on loading and unloading and on repair and rework of boards. These changes appeared to move Benchworkers away from a strictly manual assembly function to greater involvement with the automated PWBA lines – in simple tasks, such as loading and unloading, as well as on the more complex manual tasks like repair and rework.

In the two Test groups, the changes in time allocation showed an interesting polarization between the responsibilities of the hourly and salaried workers. In 1985, the hourly and salaried Test workers spent most of their time on much the same tasks: testing boards and components, and debugging PWBAs. By 1987, these tasks had become predominantly the province of the salaried testers, and the hourly Test technicians' time shifted toward inspection and repair of boards and components. This split meant that the lower-level, hourly Test jobs moved away from using the increasingly sophisticated computerized test equipment toward functions that were not machine dependent. Interviews with factory staff suggested that the hourly Test jobs were essentially simplified over this period, while the more complex testing tasks, increasingly intertwined with the computerized test equipment, were defined as the domain of the salaried testers. This type of polarization, creating a bimodal distribution of workers, to the general detriment of lesser skilled workers, was one that some scholars anticipated as an effect of computerized technology (see Denny & Fuss, 1983; Skinner & Chakraborty, 1982).

Technology Routinization

Perhaps the broadest pattern of change appeared to reflect increasing familiarity with the production technology, rather than any technology or cultural effects. For example, in the Auto-Insert group, the primary tasks in 1985 were machine monitoring, inspecting boards, and adjusting production equipment. By 1987, the time spent on all three of these activities was reduced, although monitoring and inspection remained the group's primary responsibilities. Loading and unloading PWBAs replaced adjusting equipment as the third major task of this work group. These patterns suggest that the newly computerized Auto-Insert programs were running more efficiently.

This improvement in the Auto-Insert computerized programs can be attributed to maturation of the technology as well as to improved communication between design, programming, and manufacturing personnel. In 1985, the conversion to computer software from paper tape programs had just begun. Software engineers, who developed the programs to direct the automated equipment, were largely unfamiliar with the manufacturing operations. Conversely, the manufacturing staff had limited understanding of the demands of a computerized environment. Many of the problems in the newly developed software for the Auto-Insert machines stemmed from the gaps between the engineering and manufacturing worlds. Communication bridges had to be developed between the software engineers and the manufacturing staff. During the course of implementation and development, software programmers learned more about conditions on the manufacturing floor, and manufacturing staff developed a better understanding of the specifications and controls required for computerization. As the interface with the software designers improved, the need for extensive machine monitoring and inspection was noticeably reduced. More time could then properly be spent on activities that supported the automated machine's operations.

For other work groups, noticeable shifts in time allocation across particular tasks could also be interpreted in terms of the workers' increasing familiarity with the technology. For example, the CAP group, already involved with the most extensive piece of automated technology in 1985, basically maintained the same pattern of time use. In both 1985 and 1987, the largest portions of time were spent on loading and unloading PWBAs, insertion into moving boards, inspection, and machine monitoring. The small changes that did occur in this group—an increase in the time spent on machine monitoring, and a decrease in the time spent on insertion into moving boards—imply that more of the work on the CAP PWBAs was being conducted automatically. Interviews with the factory staff suggested that this shift was probably due to more efficient machine operations as well as to improvements in matching board design to machine specifications.

Time shifts attributable to technology maturation were also found for the Maintenance and salaried Test groups. Although the Maintenance group kept basically the same task profile over time, the largest change observed was a substantial decrease in the time spent on repairing equipment—probably due to increased familiarity with the new

Table 3.7
Group- and Individual-Level Measures of Task Variety

Level and Year	*Intake*	*PAL*	*A-I*	*CAP*	*Bench-work*	*Test (Hr.)*	*Test (Sal.)*	*Main-tenance*	*Total*
Group level									
1985	4	5	3	3	4	3	2	3	3
1987	1	3	5	4	5	2	2	4	4
Diff	−3	−2	+2	+1	+1	−1	0	+1	+1
Individual level									
1985	2.4	2.2	3.1	2.3	2.0	2.0	2.6	3.0	2.4
1987	1.6	2.3	2.6	2.4	2.3	1.3	2.2	3.4	2.3
Diff	− .8	+ .1	− .5	+ .1	+ .3	− .7	− .5	+ .4	− .1
t	−2.18*	.16	− .76	.28	.80	−1.98*	−1.99*	.91	− .90

*$p \leq .10$.

equipment on the part of both the production floor and the design engineers. Similarly, the salaried Test group substantially increased the time spent on actual testing, decreasing the time spent on debugging PWBAs.

Comparisons of Task Variety at the Group and Individual Levels

Table 3.7 shows the indicators of task variety at the work group and individual levels (see the description of their computation at the beginning of this section). Two-tailed *t*-tests were conducted to assess the statistical significance of the differences between the work group means for the individual-level measures. (No statistical tests were conducted on the group-level measures because there was no within-group variance as the group-level measurements consisted of a single number for each work group.)

Table 3.7 shows that in both 1985 and 1987, somewhat higher levels of task variety were generally found at the group level than at the individual level—that is, the scope of task responsibilities of the group as a collective was slightly broader than that of the individuals within the group. This general pattern suggests some degree of specialization within the groups (i.e., individual workers tended to perform a subset of the responsibilities of the group as a whole).

In a majority of the work groups, group-level and individual-level changes in task variety occurred in parallel—where the indicator showed a narrowing of work group responsibilities, a similar narrowing of individual responsibilities was found, and vice versa. For example, task variety increased at both levels in the CAP, the Benchwork, and the Maintenance groups; it decreased at both levels in the Intake, the hourly Test, and the salaried Test groups. (Statistical tests of these changes, only possible at the individual level, showed that the changes were significant for the Intake, hourly Test, and salaried Test groups.)

The exceptions to this pattern of parallel changes on the group and individual levels were found in the Auto-Insert and PAL groups. As previously discussed, there was an increase in the number of primary tasks (from three to five) performed by the Auto-Insert group—the Auto-Insert group broadened the scope of its responsibilities. However, at the individual level, the average number of primary tasks performed by *each* Auto-Insert worker decreased somewhat, from 3.1 to 2.6 tasks. Thus, while the group as a whole moved toward performing more tasks, the individuals within the group actually focused their efforts on fewer tasks in 1987 than in 1985. The expansion of the group's responsibility was not accomplished by broadening the jobs of individual workers, but rather by assigning specialized subsets of tasks to workers in the group. Examination of the constellation of tasks performed by each individual worker in 1985 and 1987 bore out this explanation.

The PAL group showed the opposite disjuncture between the group- and individual-level indicators of task variety. At the group level, the number of "primary" tasks decreased from five to three, a change attributed to improved materials handling. However, there was a very small increase in the number of tasks performed by individual workers in the PAL group; on the average, workers performed slightly more than two tasks in both years. Again, these results could be explained in light of the combination of tasks typically performed by each worker. In 1985, performance of one insertion task was likely to be combined with one of several support tasks (e.g., loading/unloading, getting parts, or inspection); by 1987, the streamlining of the assembly process resulted in a reduction of time required for these support tasks, and most workers were able to concentrate on the two major insertion tasks.

Job Quality

A third focus of this study was the implication of the productivity improvement efforts at Factory XYZ for the quality of the jobs on the shop floor. Did the factory's goals of increasing manufacturing efficiency mean a deterioration in the quality of jobs? Were alterations in the nature and mix of tasks on the shop floor accompanied by changes in the workers' evaluations of their work experience?

Data Sources

As part of the surveys administered in 1985 and 1987, the factory production staff was asked to respond to a series of questions regarding the characteristics of their jobs. In general, these questions focused on variables that have been discussed in the literature as potentially subject to change because of the types of innovations being implemented in the XYZ factory. Most of the survey items took the form of questions or statements, with Likert-type response scales.

In this analysis, seven measures of job quality were used. *Job routinization* was a 4-item variable that measured the extent to which the work involved was repetitive. Routinization has been discussed frequently as one of the consequences of implementing automated technologies. *Personal effect* was a 2-item measure of the amount of influence that the individual perceived that (s)he had on production. *Machine effect* was a parallel 2-item measure of the perceived influence of the equipment on production. *Control* was a 2-item measure focusing on the worker's perception of control over the activities and pace of work. This variable addressed concerns that increases in automation (or, indeed, other efforts at work rationalization) result in diminishing the worker's autonomy on the job. A 4-item measure, *criticality*, addressed a similar concern – the worker's perceptions of his/her impact on the production process. A 1-item variable measured feelings of *replaceability* by advanced equipment. Finally, *overall satisfaction* was a 4-item measure inquiring about the worker's satisfaction with his/her job, supervisor, coworkers, and the company.

Table 3.8 presents the questions and Cronbach's alpha coefficients of reliability for each measure. All scores for individuals were based on the mean of the responses to each item included in the variable. To ease interpretation, all measures were converted to 7-point scales.

Table 3.8
Survey Measures of Job Quality

ROUTINIZATION—Cronbach's alpha (α) = .55

My job requires that I do the same things over and over.

(7-point scale: Strongly Disagree, Disagree, Slightly Disagree, Neither Agree nor Disagree, Slightly Agree, Agree, Strongly Agree)

I follow the same work methods or steps for doing my major tasks from day to day.

(7-point scale: Strongly Disagree, Disagree, Slightly Disagree, Neither Agree nor Disagree, Slightly Agree, Agree, Strongly Agree)

To what extent do you perform the same tasks from day to day?

1. Almost all of my tasks are the same day to day.
2. Many of my tasks are the same day to day.
3. About half of my tasks are the same day to day.
4. Some of my tasks are the same day to day.
5. Almost none of my tasks are the same day to day.

(Scale was reverse-scored and then converted to 7 points by multiplying by 7/5.)

How much the same are the day-to-day situations, problems, or issues you encounter in performing your major tasks?

1. Very much the same
2. Mostly the same
3. Quite a bit different
4. Very much different
5. Completely different

(Scale was reverse-scored and then converted to 7 points by multiplying by 7/5.)

PERSONAL EFFECT—α = .91

How much influence *do you have* on what happens in the production process?

How much influence *do you have* on what happens to the boards being produced?

(5-point scales, anchored at 1,3,5 as follows: None, Some, Very Much. Scale was converted to 7 points by multiplying by 7/5.)

MACHINE EFFECT—α = .92

How much influence *does the equipment have* on what happens in the production process?

How much influence *does the equipment have* on what happens to the boards being produced?

(5-point scales, anchored at 1,3,5 as follows: None, Some, Very Much. Scale was converted to 7 points by multiplying by 7/5.)

(continued)

Table 3.8 (continued)

CONTROL—α = .65

How much control do you have over the activities of the equipment you work with?
How much control do you have over the speed or pace at which you work?

1. No control
2. Very little control
3. Some control
4. Very much control
5. Almost total control

(Converted to 7 points by multiplying by 7/5)

CRITICALITY—α = .77

If you make a mistake in your job, how likely is it that . . .

a board would be spoiled and have to be thrown out?
the equipment/line would be shut down?
the production schedule would be seriously delayed?
the equipment would be damaged?

(7-point scales, anchored at 1,4,7 as follows: Very Unlikely; Neither Likely nor Unlikely; Very Likely)

REPLACEABILITY

How likely is it that in the next few years, machines or computers will be doing the things you now do on your job?

1. Not at all likely
2 A little likely
3. Somewhat likely
4 Very likely
(8.) I really don't know (DK)

(Converted to 7-point scale by excluding all DK [don't know] responses and multiplying by 7/4)

OVERALL SATISFACTION—α = .82

All in all, how satisfied are you . . .

with your job?
with your supervisor?
with the persons in your work area?
with (name of company), compared with other places you've worked?

(5-point scales, anchored at 1,3,5 as follows: Very Dissatisfied, Neither Dissatisfied or Satisfied, Very Satisfied. Converted to 7 points by multiplying by 7/5)

Results

Table 3.9 presents the 1985 and 1987 work group scale means and their differences. Two-tailed *t*-tests were conducted to test for the statistical significance of these differences. Because of the small number of respondents per group, a relatively lenient statistical standard, $p < .10$, was used. For the same reason, the discussion considers both statistically significant differences and overall patterns of increases and decreases over time. Table 3.10 presents the summary of directional changes by work group.

The data for the total group show mixed results for these factory workers. On the negative side, the workers reported lower levels of control on their jobs in 1987 compared to 1985. This perceived loss of control is generally in line with the increasing use of software-directed technology in the factory. These findings support the scholars and employee advocates who warn about technology's negative impacts, such as the de-skilling of jobs (Cooley, 1983) and the erosion of worker control (Shaiken et al., 1986). However, other measures show that expectations of other negative impacts were not supported. Perceptions of routinization and feelings of replaceability remained relatively stable over this period, and the decline in overall job satisfaction was nonsignificant.

The increased use of computerized technology appears to have had the effect of increasing the perceived impact of any operation—human and mechanical—on production. The production consequences of mistakes made by individuals (criticality) were reported to be more serious in 1987 than in 1985. At the same time, the workers reported greater levels of personal influence on production. Whether these perceptual changes reflected any real change in the human impact on the production process, or whether they reflected instead the impact of the quality consciousness campaign throughout the corporation, is difficult to ascertain.

It is interesting that perceptions of both person effectiveness and machine effectiveness had risen over this time period. This juxtaposition was unexpected, since we had anticipated that these perceptions would be negatively correlated. One possible interpretation is that the increasing use of computerized technology in production increases the impact of both human and mechanical operations. In a sense, the impact of any operation is multiplied by the speed with which the actions are

Table 3.9
Survey Indicators of Job Quality

Variable	Intake				PAL				Auto-Insert				CAP			
	85	*87*	*Diff*	*t*	*85*	*87*	*Diff*	*t*	*85*	*87*	*Diff*	*t*	*85*	*87*	*Diff*	*t*
N =	20	19	–	–	5	16	–	–	8	12	–	–	13	32	–	–
Routinization	5.23	4.57	–0.66	–1.95*	5.84	5.40	–0.44	– .73	5.46	5.44	–0.02	– .04	5.36	5.67	0.31	1.21
Personal effect	3.55	4.84	1.29	1.92*	5.00	4.57	–0.43	– .49	4.22	6.23	2.01	3.21*	5.41	4.83	–0.58	– .99
Machine effect	4.29	6.06	1.77	2.41*	6.17	5.29	–0.88	– .90	6.00	6.65	0.65	1.32	6.19	6.14	–0.05	– .10
Criticality	3.07	2.38	–0.69	–1.04	3.00	3.19	0.19	.19	2.02	2.27	0.25	.47	1.67	3.18	1.51	2.86*
Control	5.18	4.61	–0.57	–1.33	4.08	4.93	0.85	.94	5.12	5.12	0.00	– .01	5.74	4.52	–1.22	–2.53*
Replaceability	4.99	4.97	–0.02	– .02	6.65	4.59	–2.06	–1.94*	4.38	5.10	0.72	.85	4.71	5.47	0.76	1.11
Satisfaction	5.85	6.33	0.48	1.40	4.55	5.49	0.94	1.46	5.53	5.56	0.03	.05	6.05	5.70	–0.35	–1.00

* $p \leq .10$.

Table 3.9 (continued)

Variable	Benchwork				Test (hourly)				Test (salaried)				Maintenance				Total			
	85	*87*	*Diff*	*t*	*85*	*87*	*Diff*	*t*	*85*	*87*	*Diff*	*t*	*85*	*87*	*Diff*	*t*	*85*	*87*	*Diff*	*t*
N =	28	15	—	—	19	9	—	—	20	35	—	—	12	13	—	—	125	151	—	—
Routinization	5.19	4.98	–0.21	–.61	5.67	5.83	0.16	.52	5.04	5.48	0.44	1.96*	3.96	4.02	0.06	.15	5.20	5.30	0.10	.78
Personal effect	3.58	3.24	–0.34	–.58	2.98	3.82	0.84	1.11	3.13	3.52	0.39	.83	4.00	4.95	0.95	1.08	3.74	4.34	0.60	2.62*
Machine effect	5.05	5.20	0.15	.23	5.08	4.72	–0.36	–.39	4.23	5.09	0.86	1.97*	6.60	6.18	–0.42	–1.15	5.16	5.61	0.45	2.48*
Criticality	1.73	2.15	0.42	.89	2.23	1.66	–0.57	–1.46	1.99	2.28	0.29	.96	3.92	3.88	–0.04	–.05	2.31	2.62	0.31	1.69*
Control	5.26	4.76	–0.50	–1.29	4.84	5.48	0.64	1.36	4.76	4.72	–0.04	–.12	4.90	4.90	0.00	.00	5.05	4.78	–0.27	–1.98*
Replaceability	5.25	5.48	0.23	.32	4.51	5.44	0.93	1.19	5.08	5.15	0.07	.16	2.77	2.96	0.19	.24	4.77	5.00	0.23	.72
Satisfaction	6.00	6.00	0.00	–.01	5.46	5.94	0.48	1.12	5.52	4.80	–0.72	–2.44*	6.09	5.65	–0.44	–1.40	5.77	5.57	–0.20	–1.64

* $p \leq .10$.

Table 3.10
Summary of Changes in Job Quality Indicators

Variable	*Intake*	*PAL*	*A-I*	*CAP*	*Bench*	*Test (Hr.)*	*Test (Sal.)*	*Maint.*	*Total*
Routinization	⊖	−	−	+	−	+	⊕	+	+
Personal effect	⊕	−	⊕	−	−	+	+	+	⊕
Machine effect	⊕	−	+	−	+	−	⊕	−	⊕
Criticality	−	+	+	⊕	+	−	+	−	⊕
Control	−	+	NC	⊖	−	+	−	NC	⊖
Replaceability	−	⊖	+	+	+	+	+	+	+
Overall satisfaction	+	+	+	−	NC	+	⊖	−	−

NOTE: ○ indicates changes with $p \leq .10$.
NC indicates no change.

repeated in production and by the effectiveness of the new production tools.

As Tables 3.9 and 3.10 show, the changes in job quality indicators over time were variable across work groups. Given the differences in technology and organizational changes experienced by the different work groups over the period studied, this variability is not surprising. Brief discussions of the patterns in each work group are presented below. Caution in interpreting the results is needed, however, since some of the variability could well be due to the small number of responsibilities within a work group.

Intake

In the Intake group, perceptions of personal and machine effects on production both rose over this period, while perceptions of routinization decreased. It is likely that these shifts largely reflected the transfer of the unit out of the Intake group. This transfer involved the removal of repetitive tasks not directly linked to the production process (e.g., unloading cartons of incoming components, logging them in, and opening them) but left the preparation tasks more directly connected to the production process. The same change was probably also responsible for the slight increase in overall satisfaction.

PAL

The PAL group showed only one major shift in its job perceptions – a stability that is consistent with the lack of changes in the job or in the group's organizational structure. The main change observed was the marked decrease in feelings of replaceability, which contrasts with the increase reported by most of the other work groups. This change appears to be largely a historical artifact because, in 1985, the PAL group was feeling highly threatened by the introduction of new technology (e.g., CAP, improved Auto-Insert programs). Interviews revealed that both supervisory and nonsupervisory factory workers expected PAL to suffer most from the plans to increase automation. In fact, these fears did not come to pass; instead of layoffs, the PAL group experienced an increase in the number of jobs. The nonsignificant increase in PAL workers' reports of satisfaction is probably another indicator of the relief experienced when the dreaded cuts did not occur.

Auto-Insert

The Auto-Insert group also showed stability in most of its job perceptions. The largest change was the increase in its reports of personal effectiveness. This rise was probably due to the increased use of Auto-Insert workers on other jobs outside its work group, such as CAP and rework – functioning as "utility workers," filling in gaps as necessary in the production process.

CAP

The results from the CAP group formed an interesting pattern. The group showed a marked increase in its perceptions regarding the criticality of its mistakes for the production process, but also a decrease in perceptions of control over the pace and activities of work. Since this group was the one involved in the most highly automated process in the factory, this pattern suggests how computerized automation can affect workers' perceptions of their jobs; while personal control decreases, the consequences of one's actions become more serious. This combination of factors, if identified in other computerized settings, may provide one explanation for reports of increased stress resulting from automation.

Benchwork

Although the manual assembly function was significantly curtailed during the period studied, the nature of the activities themselves did not change in major ways. Job perceptions in the Benchwork group showed only modest changes, with no strong patterns over this period.

Test (Hourly)

Although none of the differences in the responses of the hourly Test workers were statistically significant, they hinted at some interesting contrasts. These workers appeared to feel increased levels of personal effectiveness and control, but they also showed an increase in feelings of replaceability. This apparent contradiction is reconcilable in terms of recent staffing cutbacks and shifts in task responsibilities. The hourly Test workers shifted from conducting simple tests and debugging of PWBAs to more visual inspection and repair work. These shifts made them less dependent on equipment and apparently gave them greater feelings of effectiveness. While their existing jobs may have improved subjectively, however, the large cutbacks in hourly Test positions gave rise to reasonable fears of further job losses. This group was also unusual in showing an increase in personal effectiveness along with a decrease in machine effectiveness. This pattern appears to be an effect of the group's movement *away* from automated test equipment, rather than toward its increased use.

Test (Salaried)

The increasing use by salaried Testers of computerized testing equipment seems to be reflected in their increasing feelings of machine effectiveness and perceived job routinization. This pattern is also congruent with the shift in this group's time allocation, which moved away from debugging and diagnostic work that depended on individual art, to more extensive dependence on automated test equipment. This growing reliance on automation resulted in the only significant increase in feelings of routinization in any of the work groups. This group also displayed the only significant decrease in overall satisfaction. Among the various work groups in the factory, the salaried Testers' reactions most closely resembled predictions of de-skilling and machine dominance made by employee advocates, such as Shaiken et al. (1986).

Maintenance

The results for the Maintenance group were generally stable. The most notable exceptions, although not statistically significant, were the increase in perceptions of personal effectiveness and decrease in perceptions of machine effectiveness on production.

Summary and Conclusions

This study examined the implications of productivity enhancement efforts on employment levels and on the nature of production jobs in an electronics assembly plant. The plant's initiatives included a variety of technological and organizational innovations that were so highly intertwined that it was difficult to disentangle the unique effect of any one change. However, to the extent possible, we have combined our multiple sources of data to make logical distinctions among potential explanations for the changes observed.

Previous studies have made many contradictory predictions regarding the impact of factory automation on employment and jobs. While all predictions foresee the elimination of jobs resulting from the implementation of new manufacturing technologies, the speed and extent of labor displacement are controversial. Also unclear is the extent to which new forms of employment will be generated as a direct result of the technologies themselves, or as secondary effects of general growth facilitated by technology. In general, earlier predictions, which have been rather global and deterministic in character, have given way to forecasts that include more contingencies and multiple possibilities.

Our intensive study of the experiences in one electronic assembly plant supports the predictions of complex and variable effects. In general, the study showed that the impacts of factory automation efforts can vary substantially even across work groups within a single plant. Thus while factory automation did result in a reduction of the overall employment level in the factory, as expected, the sizes of individual work groups shifted in both directions. As might be expected, the work groups involved in manual assembly suffered major reductions, while those involved in more automated processes increased in size. Variability across work groups was also seen in terms of job structure and job characteristics. While some of the jobs narrowed in scope (e.g., Intake and PAL jobs), others broadened (e.g., Auto-Insert jobs), and yet others were substantially redefined (e.g., Benchwork and Test jobs).

Some of these changes could clearly be attributed to technology. However, our study also demonstrated the important roles that nontechnological factors, such as management philosophies, played in affecting how the reduction in employment levels was experienced by individual employees. In Factory XYZ, as in many other plants embarking on the implementation of new technologies, the factory automation effort was, in large part, motivated by a goal of reducing human labor in the production process. Although the change could have been accomplished in a manner that was highly traumatic to the employees, the amount of actual displacement suffered by individual workers was minimized by the company's incremental approach to new technology implementation, its careful personnel policies, and continued dialogue with the employee union.

It is clear that the impacts of factory automation efforts cannot be usefully studied exclusively in terms of the new hardware and software being installed on the plant floor. Defining the automation effort in such narrowly technological terms would reflect reality rather poorly. As this case shows, new manufacturing technologies are generally implemented, not in a vacuum, but as part of a broad organizational response to perceived needs for improvement. Further, the installation of new production technologies was used by the organization as the occasion for rationalizing established practices, work rules, and structures. While some changes (e.g., the decrease in the number of manual assembly jobs coupled with the increase in automated assembly jobs) could be directly attributed to the new technologies, other changes were more attributable to changes in production philosophy and managerial priorities made within the factory and at the corporate level. The mingling of technological and other organizational innovations, which makes it impossible to examine the effects of factory automation in purely technological terms, is probably a key to the successful implementation of such efforts. For instance, the Manufacturing Studies Board (1986) suggested that the implementation of advanced manufacturing technologies is most effective when accompanied by philosophy grounded in improving the plant culture, management, organizational and job structures, and labor-management relations.

Epilogue

It is useful to remember the importance of the time frame in any study of changes accompanying factory automation. This study documented employment and job changes over a three-year period in the middle-1980s that could be considered an early phase in the push toward becoming a "factory of the future." Many of the reported changes in job structure may have reflected transitional adjustments to a still maturing technology, rather than more permanent alterations in the "final" job structure at the plant. And yet, perhaps the job structure at the factory may remain in transition for a long time. Although the PWBA production process has not basically changed, the move toward increasing automation continues incrementally. At this writing, three years after the end of the study, there are eight automated assembly lines instead of two; a new automated storage and retrieval system (ASRS) is in place. Production levels at the plant have not changed substantially, although even more board varieties are now being produced. The work is being performed by an even smaller work force (about 220 employees).

The company has made substantial gains in its campaign to regain its former position as world leader. Perhaps a key change in terminology can provide a clue as to whether the company is likely to achieve that goal. As an indication of the extent to which the increased emphasis on quality has become part of the plant culture, production errors are no longer counted in terms of "percent defective"—parts per hundred that are defective—but rather errors are tallied in "parts per million defective."

References

Andelin, J. (1983, March). *Manufacturing automation: Selected employment and education issues* (Testimony before the Joint Economic Committee, U.S. Congress). Washington, DC: Office of Technology Assessment.

Argoff, N. J. (1984, May). Selected education and training case studies. In *Computerized manufacturing automation: Employment, education and the workplace* (Vol. 1, App. A). (OTA-CIT-235). Washington, DC: Office of Technology Assessment.

Ayres, R. U., & Miller, S. M. (1982, May/June). Industrial robots on the line. *Technology Review*, pp. 35-46.

Ayres, R. U., & Miller, S. M. (1983). *Robotics: Applications and social implications.* New York: Harper & Row.

Bednarzik, R. W., & Schoepfle, G. K. (1985, December). *Technology and employment in the United States: Past, present, and future.* Paper presented at the International Symposium on Microelectronics and Labor, Tokyo.

Blauner, R. (1964). *Alienation and freedom.* Chicago: University of Chicago Press.

Braverman, H. (1974). *Labor and monopoly capital.* New York: Monthly Review Press.

Cooley, M. (1983). *Architect or bee?* Boston: South End Press.

Cronbach, L. J., & Gleser, G. C. (1953). Assessing similarity between profiles. *Psychological Bulletin, 50,* 456-473.

Denny, M., & Fuss, M. (1983). The effect of factor prices or technological change on the occupational demand for labor: Evidence from Canadian telecommunications. *Journal of Human Resources, 17,* 161-176.

Hackman, J. R., & Oldham, G. R. (1974). *The job diagnostic survey* (Tech. Rep. No. 4). New Haven, CT: Yale University, Department of Administrative Sciences.

Hartmann, G., Nicholas, I., Sorge, A., & Warner, M. (1984). Computerised machine-tools, manpower consequences, and skill utilisation: A study of British and West German manufacturing firms. *British Journal of Industrial Relations, 21,* 221-231.

Hazelhurst, R. J., Bradbury, R. J., & Corlett, E. N. (1969). A comparison of the skills of machinists on numerically controlled and conventional machines. *Occupational Psychology, 43,* 169-182.

Hunt, H. A., & Hunt, T. L. (1983, March). *Human resource implications of robotics.* Kalamazoo, MI: Upjohn Institute for Employment Research.

Kohler, C., & Schultz-Wilde, R. (1983). *Flexible manufacturing systems: Manpower problems and policies.* Paper presented at World Congress on the Human Aspects of Automation, London.

Kutscher, R. E. (1984, February). *Future market conditions for engineers.* Paper presented at the National Research Council Symposium on Labor Market Conditions for Engineers, Washington, DC.

Leontief, W., & Duchin, F. (1984, April). *The impacts of automation on employment, 1963-2000* (Final Report to NSF Contract No. PRA-8012844). New York: New York University, Institute for Economic Analysis.

Levin, H. M., & Rumberger, R. W. (1983, January). *The educational implications of high technology* (Project Report No. 83-A4). Stanford, CA: Stanford University, Institute for Research on Education.

Lublin, J. (1981, October 28). Steel collar jobs: As the robot age arrives, labor seeks protection against loss of work. *The Wall Street Journal,* p. 1.

Majchrzak, A., Chang, T. C., Barfield, W., Eberts, R. E., & Salvendy, G. (1987). *Human aspects of computer-aided design.* London: Taylor & Francis.

Majchrzak, A., Nieva, V. F., & Newman, P. D. (1986). Adoption and use of computerized manufacturing technology: A national survey. In D. D. Davis (Ed.), *Managing technological innovation* (pp. 105-126). San Francisco: Jossey-Bass.

Manufacturing Studies Board, Committee on the Effective Implementation of Advanced Manufacturing Technology, National Research Council. (1986). *Human resource practices for implementing advanced manufacturing technology.* Washington, DC: National Academy Press.

Nieva, V. F., Gaertner, G. H., Argoff, N. J., & Newman, P. D. (1986). *Work in a changing factory* (Interim Report to the National Science Foundation), Rockville, MD: Westat.

Office of Technology Assessment. (1984, February). *Computerized manufacturing automation.* Washington, DC: U.S. Government Printing Office.

Piore, M. J., & Sabel, C. F. (1984). *The second industry divide.* New York: Basic Books.

Shaiken, H., Herzenburg, S., & Kuhn, S. (1986). The work process under more flexible production. *Industrial Relations, 25*, 167-182.

Skinner, W., & Chakraborty, K. (1982). *The impact of new technology: People and organizations in manufacturing and allied industries.* Elmsford, NY: Pergamon.

Taylor, J. C., & Bowers, D. G. (1972). *The survey of organizations.* Ann Arbor, MI: Institute for Social Research.

U.S. Department of Commerce, Bureau of the Census. (1988). *U.S. industrial outlook 1988—Electronic components.* Washington, DC: U.S. Government Printing Office.

Van de Ven, A. H., & Ferry, D. L. (1980). *Measuring and assessing organizations.* New York: John Wiley.

Warner, M. (1984). New technology, work organizations and industrial relations. *Omega, 12*, 203-210.

PART II

OFFICES

4

Computer Use, Control Over Computers, and Job Satisfaction

BARBARA A. GUTEK
SUSAN J. WINTER

The drive to computerize white-collar work is motivated by a desire to raise white-collar productivity, which is considered notoriously low relative to other types of work, such as farming and manufacturing (Shaiken, 1984; Strassman, 1985). The diffusion of computer equipment into white-collar work environments is picking up momentum, facilitated by declining hardware costs and increasingly sophisticated, diverse, and "user friendly" software. There is evidence, for example, that the rate of diffusion of personal computers (PCs) has been increasing within the past 5 to 6 years (McPadden, 1988). Declining hardware costs have been a particularly strong force in the computerization of work. According to Smith (1984, cited in Kraut, 1987), 1980 marked the turning point when an hour of computer time became cheaper than an hour of a person's time to accomplish the same task.

The impact of this proliferation of computer equipment and computerized procedures on workers and work groups is therefore an impor-

---

AUTHORS' NOTE: This research was partially supported by Grant Number IRI-8714768 from the National Science Foundation, Information Impacts Program, to Barbara A. Gutek.

tant topic of social science research. Although social scientists have not come forward in large numbers to study computers as a social and psychological phenomenon, there are clusters of researchers who are engaged in systematic, programmatic research. (See, for example, research by Kiesler, Sproull, and colleagues at Carnegie-Mellon University, Majchrzak and colleagues at USC, Kraut and colleagues at Bellcore, and Bikson and colleagues at the RAND Corporation.) Some researchers have directly tested social science theories relevant to computer use (see also Griffith, 1988; Kiesler, Siegel, & McGuire, 1984; Sproull, Kiesler, & Zubrow, 1984), whereas the programs of research conducted by others have been drawn from real-world concerns or from a combination of theory and the concerns of computer vendors, implementers, and users (e.g., Majchrzak, 1988). Our interests in this topic grew out of experiences with powerful computer systems in office applications, and led to studies of the implementation of information systems in work groups (Bikson & Gutek, 1983; Bikson, Gutek, & Mankin, 1987; Gutek, 1983; Gutek, Bikson, & Mankin, 1984).

In this chapter, we explore the effect of computerization of work on job satisfaction of the employees, focusing particularly on some possible intervening reactions of workers to computer use. Previous attempts to show consistent direct positive or negative effects of computer use have been largely unsuccessful (for a review and discussion, see Gutek et al., 1984; Kraut, 1987). Thus, the relationship between computer use and outcomes may be mediated by people's perceptions of and reactions to the technology.

Framework of This Research

We begin with a few assumptions. First, we assume that computer systems are sets of loosely bundled technologies that are highly adaptable to many uses, styles of use, work flows, and organizational structures (Bikson, 1987; Kraut, Egido, & Galegher, in press; Rousseau, 1986). By loosely bundled technologies, we mean that computers combine a number of different technologies and come in many shapes and forms. Here we focus on a broad subset of computers, namely any central processing unit (CPU) with at least a keyboard (an input device) and a monitor that allows the user to see what is happening (an output device). Because users frequently want a relatively permanent copy of their output, most systems are also connected to one or more printers that provide "hard-copy" output.

Second, we assume computer systems are sufficiently flexible that managers can implement their use under widely diverse management styles; for example, a specific computer system can be implemented and used in a highly regimented autocratic style or in a laissez-faire manner (Bikson et al., 1987; Strassman, 1985). The technology does not narrowly determine the conditions under which it is used (Gutek et al., 1984); in other words, there is relatively little technological determinism.[1]

Third, we assume that what is important to workers about a computer system is not its specific hardware configuration but the functions for which it is used (Bikson et al., 1987). A mainframe computer can be used for communication, as can a microcomputer. A dedicated word processor is usually used for word processing, but so are many microcomputers, and some people do word processing on a mini or mainframe.

In this chapter, we describe the current status of computer-using work groups with respect to the computer tools available to them, workers' experience with and knowledge of computer systems, and workers' satisfaction with their jobs. We also describe the extent to which computer surveillance and personnel surveillance are used in the work group. In addition, we explore two intervening psychological mechanisms by which computer systems and computer use can affect a worker's job satisfaction: symbiotic interaction and perceived control.

Symbiotic interaction is the degree to which the worker feels that he or she and the computer work well together. Computers represent an alien culture for new users (Sproull et al., 1984), and computer anxiety has replaced math anxiety as the most recently discovered, widespread personal fear. Computer anxiety is expected to affect a person's attitude toward a computer as well as his performance. People who are high in computer anxiety may be especially unlikely to like or be satisfied with computers. Students who had little interest in computers or math and did not like computers were found to be less likely to become computer literate than others (Miura, 1984; see also Seidel, Anderson, & Hunter, 1982). People who feel that their computer tools allow them to work smoothly and efficiently may be more satisfied than those who feel that they and their computer system are unproductive or slow. In the latter case, workers may feel frustrated in their interaction with the specific tool, and lower job satisfaction may result. Thus, we predicted that a smooth, efficient interaction between the end-user (the computer-using employee) and the computer system would be associated with high job satisfaction.

The second intervening psychological mechanism we studied was *perceived control* over the computer. People attempt to understand their world (Heider, 1958; Kelley, 1973) in order to establish a sense of control. Having a sense of control over one's life generally has positive outcomes for the individual (e.g., Cohen, 1978; Langer & Saegert, 1977; Schulz, 1976). Thus, we predicted that workers who felt they controlled their use of the computer and believed that the computer was a tool for their use would be more satisfied with their jobs than users who felt they had little choice if, when, and how they used a computer system or who felt their job was to use the computer (Gutek, 1983; Kling, Iacono, & George, 1989).

The issue of control is especially relevant for the study of computers because of the computer's ability to monitor the user's performance. Computers excel at manipulating data and those data can include information as to how computer operators use them. Several writers and researchers (Attewell, 1987; Kling & Iacono, 1984; Zuboff, 1988) have expressed concern that computers will become a tool of surveillance, replacing or supplementing managerial supervision. (Although it is widely known that AT&T uses a computer-monitoring system with its telephone operators, relatively little other evidence for widespread surveillance yet exists—Attewell, 1987.) Thus, as part of our interest in the use of computers and perceived control, we describe the extent to which workers in computer-using groups were subject to monitoring of their work, both by their supervisor and the computer.

The framework we used for exploring and describing the relationships among computer use, psychological mechanisms, and job satisfaction is shown in Table 4.1. We explored two aspects of workers' computer use: the available computer tools and the skill and experience of workers with computers. We expected that more computer tools and greater computer skill and experience would be associated with both a more symbiotic interaction with computers and greater perceived control over them. We expected both the adequacy of computer tools and worker skill and experience with computers, as well as the two psychological mechanisms, to be associated with job satisfaction.

Table 4.1
A Framework for Studying Computer Use and Job Satisfaction

Category of Variables	*Variables*
Computer use	Tools Skills and experience Monitoring/surveillance
Intervening psychological mechanisms	Symbiotic interaction Control over the computer
Control variable	Routineness of work
Outcome variable	Job satisfaction

Many other factors might affect or confound the relationships among computer tools, psychological mechanisms, and job satisfaction, notably characteristics of one's job that directly affect job satisfaction (see Hackman & Oldham, 1975; Quinn & Cobb, 1980). The routineness of the job is one characteristic that has been implicated in research and theoretical discussions concerning computerization of work (Braverman, 1974; Feldberg & Glenn, 1987; Working Women Institute, 1980). People who do very routine jobs may be expected to use computers in a routine (i.e., short-cycle, repetitious, low skill) manner, which is likely to affect their level of job satisfaction. It may be the routineness of the job rather than the computer, per se, that affects job satisfaction. Thus, we included routineness of work as a control variable in order to assess the effects of computer tools and skills, symbiotic interaction, and perceived control over the computer on job satisfaction, unconfounded by the routineness of the job.

In summary, we investigated the relationships among computer use variables, reactions to computers, and reactions to the job, as well as the amount of monitoring and surveillance by managers and computers. As intervening variables, we measured the perceptions of workers as to how well they and the computer work together (symbiotic interaction) and their perceptions of control over the computer system; we then related these factors, as well as the routineness of the job, to job satisfaction. Our intent was to provide descriptive information about some aspects of the computerization of work, about which there has been speculation based mostly on personal experience, observation, or case study (Shaiken, 1984; Strassman, 1985; Zuboff, 1988), but little systematic empirical evidence.

Method

Data for this report came from a study of computer-using white-collar work groups. Consistent with previous research (Bikson & Gutek, 1983; Bikson et al., 1987; Gutek et al., 1984), we defined a work group as four or more people, including at least one level of supervisors, engaged in some common information-related process or product. We wanted participants who were diverse in computer implementation and use as well as in computer configurations.

Work groups were eligible for the present study if computers had been used by the group for at least one year and if computers were necessary for the work of the *group* (rather than for the work of one or two individual employees) though not all people in the group had to personally use a computer. In this study we collected questionnaire data from all members of 89 computer-using work groups in 49 different organizations in the Southern California area, and we interviewed the supervisor or manager of each work group in the summer of 1988. The semistructured interviews took about 90 to 120 minutes each. Nine to 10 months later, each supervisor was reinterviewed for 30 to 40 minutes by telephone; 88 supervisors were reinterviewed, because one work group no longer existed. The present analyses are based on the responses of 89 supervisors in interviews and 623 questionnaire respondents in the 89 work groups.

Sample

In examining the 89 work groups, we found that 26% were from companies primarily engaged in manufacturing, and 74% were from companies primarily providing services. Of the work groups, 45% were in the public sector, with 55% in the private sector; 53% were performing line functions, while 47% were staff. When classified according to the nature of their work, using a four-category classification developed in previous research (Bikson & Gutek, 1983; Bikson et al., 1987), the work groups were distributed as follows: 22% managerial or administrative units (e.g., finance, city manager's office); 25% text-oriented professional units (e.g., specialized library department, organizational effectiveness, marketing); 26% technical professional units (e.g., computer-related research department, industrial engineering, computer-aided design); and 27% clerical units (e.g., auto parts department, billing department, dispatch unit, order-taking department). The number of workers per group ranged from 4 to 39 and showed a positive skew with a median of 9 and a mean of 10.1.

The relatively recent influx of computer systems into all types of white-collar work suggests that many workers are relative novices as computer end-users. Our data are illustrative; the average work group in this study first started using computers in 1984. In addition, in the intervening 8 to 10 months between the first and second data collection, 46% of the work groups had added to the number of personal computers (PCs) available to them.

Because we were interested in psychological mechanisms that intervene between computer systems, user skill, and experience on the one hand and job satisfaction on the other, we focused on the 623 individuals in this study rather than on entire work groups. Of the employees, 52% had professional or managerial positions, while 48% worked at technical, clerical, or secretarial jobs; women comprised 64% of the sample. Only 5% of the employees earned less than $10,000 a year, 40% earned between $10,000 and $25,000, 45% earned between $25,000 and $50,000, and 9% earned over $50,000. Individuals in our sample were well educated: 4% had doctoral degrees, 21% had master's degrees, 18% had bachelor's degrees, 38% had some college, 6% had attended vocational school, and 13% had completed high school. Respondents were also fairly young: 12% were under 25 years old, 39% were 26 to 35, 28% were 36 to 45, 17% were 46 to 55, and 6% were over 55. Regarding marital status, 45% of the sample were currently unmarried and 55% were married.

Measures

The antecedent variables of computer use, availability of computer tools, computer skills, and the topic of monitoring of work and computer surveillance were measured by individual questions designed for this study. These questions and their descriptive findings are presented in the Results section of this chapter. Two intervening psychological mechanisms of symbiotic interaction and perceived control were also measured by a series of questions designed for this study. These questions were subsequently factor-analyzed and scales were created (see below). The dependent variable (job satisfaction) and the control variable (routineness of work) were multiple-item scales that had been constructed and used in previous research.

Psychological Mechanisms

Symbiotic interaction was measured with five semantic differential items created for this study. Respondents were asked to describe "the computer and me working together" on five items (inefficient-efficient, rough-smooth, unproductive-productive, useless-useful, slow-fast). These items, each measured on a five-point scale, formed an internally consistent scale (alpha = .83) that was highly skewed toward the negative end (Mean = 4.15, *SD* = .77). That is, most of the workers in our study claimed that they and the computer worked very well together.

Perceived control over the computer was also measured on five-point scales by five semantic differential items shown in Table 4.2. Four of the items formed a reasonably reliable scale (alpha = .57) and were combined into a variable called "control over the computer." The 4-item measure was highly skewed toward the negative end (Mean = 4.06, *SD* = .83), with most people reporting high perceived control over the computer. The fifth semantic differential item—"the computer is a tool to be used at my convenience" versus "my job is to use the computer to do my work"—was not related to the other four items and so was treated separately. The distribution of responses to this item was slightly U-shaped, with the largest numbers of employees at the two extremes, asserting either that the computer was a tool to be used at their convenience *or* that their job was to use the computer.

Dependent Variable

Job satisfaction was measured by the 2-item facet-free job satisfaction measure taken from the University of Michigan Quality of Employment Survey (Quinn & Staines, 1979). The items were: "All in all, how satisfied would you say you are with your job? Very satisfied, somewhat

Table 4.2
Measures of Control Over the Computer (a 4-Item Scale and a Single-Item Variable)

	Anchor (1)	*Anchor (5)*
1.	The computer uses me.	I use the computer.
2.	I cannot choose when to use the computer.	I can choose when to use the computer.
3.	The computer controls me.	I control the computer.
4.	I am here to use the computer.	The computer is here to assist me. (alpha for four-item scale = .57)
	My job is to use the computer to do my work.	The computer is a tool to be used at my convenience.

satisfied, not too satisfied, not at all satisfied;" and "Knowing what you now know, if you had to decide all over again whether to take the job you now have, what would you decide? Would you decide without any hesitation to take the same job? Would you have some second thoughts? Would you decide definitely not to take the same job?" This 2-item, reliable scale (alpha = .72; range = 1 to 3.5) is skewed toward the negative end (Mean = 2.93, *SD* = .60); that is, most respondents were satisfied with their jobs.

Control Variable

Routineness of work was measured by the Withey, Daft, and Cooper (1983) 5-item indicator of routinization. The five items had a high internal consistency (*N* = 611, alpha = .88) and were combined into a scale that was approximately normally distributed (Mean = 2.45, *SD* = .72).

Descriptive Results

Computerization of Work

The extent of computerization of work was assessed three ways. First was the percentage of people in the sample of computer-using groups who said that they personally used a computer in their work. Of the 617 people who answered the question, 95% said that they personally used a computer in their current job.

Second, we asked people if they shared a computer with others or had their own terminal. This is an important point because computer use is limited by one's access to a computer—that is, workers who share a computer with one or more people cannot use it as often as those who have their own computer terminal. As Table 4.3 shows, slightly more than half of the workers had their own terminal, with 5.7% saying they had more than one terminal of their own. Most of the people who shared a terminal shared with more than one other person (44% of the total group). In most cases of sharing, computer terminals were available in a particular location that was accessed by all or most members of the work group—and occasionally by members of other work groups as well. In some cases, the computer area contained several of the same pieces of equipment; in other cases, it contained a mix of systems (for example, a dumb terminal hooked to a mainframe, a PC with a dot matrix printer, and another PC connected to a laser printer).

The third measure of computerization was the ratio of workers per terminal in the work group. In the 89 work groups, the ratio ranges from .47 (less than half a person per terminal or more than two terminals per worker) to 15 workers per terminal. The median score for the 89 work groups was 1.25 workers per terminal; 20% of the work groups had exactly one worker per terminal, while about 24% had less than one person per terminal.

Computer Skills and Experience

The computer skills and experience of workers in the study were measured in several ways. First was the number of hours per week that the worker spent at the computer. As Table 4.3 shows, time spent using a computer yielded a fairly flat distribution, with over 30% of users reporting light usage of computers (less than 10 hours a week) and about 25% reporting heavy usage (over 30 hours a week).

A second measure used was the number of years of computer experience. Unlike the flat distribution of hours per week, the number of years of computer experience was closer to a normal distribution. There were 4% of the workers who said they had no computer experience, whereas 11% said they had more than 10 years of experience. The modal category of experience was between one and five years (see

Table 4.3
Measures of Computer Resources, Experience, and Skill

Question and Responses	*Number*	*Percent*
Do you share a terminal? (N = 560)		
Share with more than one person	248	44.3
Share with one person	24	4.3
Have one terminal	256	45.7
Have more than one terminal	32	5.7
Hours per week at the computer (N = 567)		
Less than 5	82	14.5
5-10	101	17.8
10-20	112	19.7
20-30	126	22.2
30-40	112	19.7
More than 40	34	6.0
Years of computer experience (N = 616)		
None	25	4.0
Less than 1 year	63	10.2
1-5 years	281	45.6
5-10 years	177	28.7
More than 10 years	70	11.4
Respondents' stated computer skills (N = 618)		
Data entry	455	73.6
Applications (e.g., word processing)	367	59.4
Programming language(s) (e.g., BASIC)	155	25.1
Machine language	53	8.6
Hardware/systems	6	1.0

Table 4.3). These results reflect the relatively recent sharp increase in computer use in white-collar work and are consistent with our findings for work groups (i.e., computers were first introduced into the average work group in the study in 1984, 4 years prior to the beginning of our research).

In addition to computer experience, we also asked about workers' computer skills. Table 4.3 shows the percentage reporting the following computer skills: data entry, applications packages, programming, machine language, and hardware or system skills. In general, the lower the level of the skill, the more people reported they had that skill. Almost 75% of the workers said they had data entry skills, and 60% could run an application program like a word processing program, spreadsheet program, or statistical package. However, only 25% reported some programming skills, and less than 10% said they knew machine language. When the total number of computer skills was calculated, the modal category was one, and 77 workers (12%) said they had none of the computer skills listed. Only 5 percent of the sample had four of the skills, and no one had more than four.

These reports of computer skills seemed somewhat low to us, but supervisors of the groups did not necessarily agree that their work group had a relatively low level of skill. In the second interview, we asked each supervisor what proportion of their work group had "a solid, basic understanding" of their computer system. Almost one-quarter of the supervisors said everyone had such an understanding, 30% claimed that over half did, and only 2% said no one in their work group had a solid, basic understanding of their computer system.

Computer Surveillance

Another aspect of computer use that has been the subject of speculation and not much research is the use of computers for surveillance or monitoring of workers. We asked employees about the monitoring of their work on the questionnaire (summer 1988) and asked supervisors about the monitoring of employees' work in the second interview (summer 1989). Table 4.4 shows what workers said their supervisor knew and what the supervisor claimed to know about the workers in his or her group, whether by personal or computer monitoring. The data indicated that while a majority of supervisors knew the general behavior of employees, such as the number of hours on the computer, far fewer knew detailed behavior, such as number of keystrokes or transactions. In general, supervisors reported more surveillance or monitoring than employees. For example, 48% of the employees said their supervisor knew how many hours a day they spent at the computer, whereas 57% of the supervisors said they usually knew; an additional 28% of the supervisors said they sometimes knew. (Note that the figures for hours per day at the computer and number of applications used may not be good indicators of surveillance because some employees spend 8 hours a day at the computer and use only one application—thus, the

Table 4.4
Responses by Employees and Supervisors About Amount of Surveillance

Item	*Workers (N = 560)* *% Yes*	*Supervisors (N = 88)* *% Sometimes*	*% Usually*
Supervisor knows how many hours a day on computer	47.9	28	57
Supervisor knows number of applications used	54.1	28	58
Supervisor knows how many errors or mistakes made on computer	16.3	25	22
Supervisor knows number of keystrokes or transactions	11.1	16	16

supervisor would know such information without any surveillance at all.) The figures for more detailed information—number of errors or mistakes on the computer and number of keystrokes provided more precise indications of computer monitoring. Here the figures provided by employees suggested that 11% to 16% of the employees were monitored by computer, whereas figures provided by supervisors suggested that workers in 32% to 47% of the work groups were monitored by computer.

Computer surveillance makes possible a more detailed level of surveillance. At the least detailed level, 85% of the supervisors claimed they sometimes or usually knew the number of hours per day workers were on the computer, but in only 18 of the 75 groups (24%) did this information come from the computer. Similarly, 86% of the supervisors knew how many applications programs their workers used, but in only 22 of 76 groups (29%) did the information come from the computer. About 47% of the supervisors claimed to know at least sometimes the number of errors their workers made, and about half of the time (22 of 41 groups) this information came from the computer. Regarding the most detailed information—number of keystrokes or transactions—only 32% of the supervisors claimed to have this information, and in 82% of the cases (23 of 28 work groups) this information was provided by the computer.

Table 4.5
Percent of Supervisors Reporting Computer-Generated Information on Productivity

Productivity Measure	*Total Sample (N = 89)*	*Managerial Groups (N = 20)*	*Text Professional Groups (N = 22)*	*Technical Professional Groups (N = 23)*	*Clerical Support Groups (N = 24)*
Employee productivity	31	20	32	17	54
Group productivity	33	15	23	43	46

Altogether, 28 supervisors (31%) reported that the computer had generated information about the individual performance of some or all of the employees (see Table 4.5), but 3 of the 28 supervisors (11%) said that the information was not used. About 33% said that the computer generated information about group performance, and again three supervisors said that the computer-generated information was not used—either because the information was not presented in a useful form or because the supervisors simply did not need it.

Not surprisingly, supervisors of clerical work groups were the most likely to report computer-generated information about some or all of the workers' performance (Table 4.5). While 54% of the supervisors of clerical offices reported having computer-generated performance data, only 17% of the managers of the technical professional work groups received individual performance data via computer. In contrast, for *group* performance data, surveillance was most likely in both clerical and technical professional work groups (about 45% in both categories). At the low end of surveillance were managerial work groups; only 15% of the supervisors of these groups received any computer-generated information about group productivity.

Relationships Among Variables

In this section, we explore some relationships among computer experience and skills, and between computerization and the psycholog-

Table 4.6
Relation of Computer Skills to Experience With Computers

Skill	*Mean experience*[a]	*% not using a computer at home*	*N*
Data entry	3.47	61	454
Applications packages (e.g., spread sheet)	3.51	50	365
Programming language(s) (e.g., Pascal, BASIC)	3.73	41	154
Machine language	3.85	31	52
Hardware/systems	4.17	17	6

[a]1 = no experience; 2 = <1 yr; 3 = 1-5 yrs; 4 = 5-10 yrs; 5 = >10 yrs.

ical mechanisms of symbiotic interaction and perceived control. The relationships between the two psychological mechanisms and job satisfaction will then be examined. Finally, we regress job satisfaction on the aforementioned measures, with routinization of work serving as a control variable.

Computer Use and Skill

While 95% of the workers in the 89 work groups claimed to use a computer in their work, 12% reported that they had none of the computer skills about which we inquired. This is possible because a person who only accesses information by computer, for example, would not require any of the skills shown in Table 4.6.

We might expect that the workers who had the most previous experience with computers would also have more computer skills than those who were recently initiated to computer use. Table 4.6 shows the relationship between computer skills and two aspects of previous computer use: length of past computer experience, and the use of computers at home. In general, people who had more advanced skills, such as programming or machine language, had more experience with computers than those who had lower level skills. Table 4.6 also shows that people who had more advanced skills were more likely to use a

Table 4.7
Correlations Between Computer Experience and Skill and Intervening Psychological Mechanisms

Measures of Experience and Skill	*Symbiotic Interaction*	*Control Over Computer*	*Computer as a Tool*
Hours per week at computer	.06 (519)	.01 (521)	–.08 (520)
Years of experience with computer	.22** (556)	.02 (559)	–.07 (557)
Computer skills	–.18** (556)	–.18** (559)	–.12* (557)

NOTE: *N*s are in parentheses.
*$p < .01$; **$p < .001$.

computer at home than their less skilled counterparts. For example, 61% of the people who said they had data entry skills said they did not use a computer (either their own or the company's) at home (either for work or personal use), whereas only 31% of the people who knew machine language and only one of the six people who had some knowledge of computer hardware did not use a computer at home.

Computerization and Psychological Mechanisms

The two intervening psychological mechanisms examined in our analyses, symbiotic interaction between computer and end-user and perceived control over the computer, were not entirely independent. A smooth, fast, efficient, productive interaction with the computer was associated with the end-user's perception of greater control over the computer ($r = .30$, $N = 554$, $p < .001$) and was weakly but reliably associated with the perception that the computer was a tool to be used at the end-user's convenience ($r = .18$, $N = 553$, $p < .001$).

Symbiotic interaction was related to computer skill and experience, but not in a consistent manner (see Table 4.7). The perception that the worker and computer work well together was unrelated to the number of hours per week the user spent at the computer, positively related to the number of years of experience with computers ($r = .22$, $N = 556$, $p < .001$) and negatively related to computer skill level

($r = -.18, N = 556, p < .001$). The last result, that people who had more computer skills reported a *less* symbiotic relationship with the computer, was surprising. Possibly the workers who did relatively narrowly prescribed computer tasks also had fewer computer skills but worked well with the computer at their narrow tasks. For example, some people in the survey used the computer simply to call up one or two screens and read information from the screen. Very little computer skill is called for in such a task, and as long as the computer responds reasonably quickly when the worker tries to access a particular screen, the worker might consider that he or she and the computer worked well together. In contrast, the person who had much more computer expertise but was using the computer (or several computers) for many more varied and complex tasks might experience many more problems and frustrations and consequently might feel that he or she and the computer did not work especially well together.

Perceived control over the computer was the second intervening psychological mechanism, measured by both a 4-item scale and a single semantic differential item (shown in Table 4.2). In general, the use of computers and experience with computers were *not* related to a perception of control over the computer (see Table 4.7). People who worked many hours a week with computers or who had many years of experience with computers did not report any more control over them than workers with little experience. However, skill level was weakly and negatively associated with perceived control over the computer ($r = -.18, N = 559, p < .001$ for control over computer scale, and $r = -.12, N = 557, p < .01$ for computer as a tool). In both cases, people with higher skills perceived they had *less* control over the computer than people with relatively lower skills.

Psychological Mechanisms and Job Satisfaction

Neither a symbiotic interaction between the computer and the end-user ($r = .09, p < .01$) nor control over the computer was strongly related to job satisfaction ($r = .14, p < .001$ for control over computer, and $r = .10, p < .01$ for computer as a tool), although results were significant because of the large sample ($N \geq 547$—see Table 4.8). Interaction between the computer and the end-user was unrelated to the control variable, routineness of work, but perceived control over the computer was related to routineness of work. People who perceived that they had control over the computer had less routine jobs ($r = -.23, p < .001$), and

Table 4.8
Correlations Between Intervening Psychological Mechanisms and Outcome and Control Variables

Intervening Psychological Mechanisms	*Job Satisfaction*	*Routineness*
Symbiotic interaction	.09* (547)	.03 (551)
Control over computer	.14** (551)	–.23** (551)
Computer as a tool	.10* (549)	–.35** (549)

NOTE: *N*s are in parentheses.
*$p < .01$; **$p < .001$.

people who reported that the computer was a tool to be used at their convenience had considerably less routine jobs than the other workers ($r = -.35$, $p < .001$).

Table 4.9 shows the results of a hierarchical multiple regression analysis in which job satisfaction was regressed on the control variable (routineness of work), on the intervening psychological mechanisms of perceived control and symbiotic interaction, and on the computer skill and experience variables and characteristics of computer tools available to workers. The set of predictors together yielded a multiple R of .33 (adjusted $R^2 = .09$). Table 4.9 shows that the strongest predictor of job satisfaction, by far, was low routineness of work. Routineness of the job itself yielded an adjusted $R^2 = .07$. When routineness of the job was controlled, perceived control over the computer system was not related to job satisfaction, but symbiotic interaction was a weak predictor of job satisfaction ($F = 6.1$, $p < .01$). End-users who felt they and the computer worked well together were slightly more satisfied with their jobs than end-users who felt they did not work particularly well with their computer system.

The only aspect of the computer system predictive of job satisfaction was years of experience with the computer ($F = 5.5$, $p < .05$), and the direction of the relationship was positive: People with more computer experience reported slightly higher levels of job satisfaction than their less experienced counterparts. Sharing a terminal or having a small ratio of workers to terminals was not related to job satisfaction.

Table 4.9
Predictors of Job Satisfaction

Variables	*Final Beta Weight*	*F-ratio of Beta Weight*
Routineness of work	–.27	29.2***
Control over computer	–.06	1.4
Computer as a tool	–.04	.9
Symbiotic interaction	.12	6.1**
Hours/week at computer	.02	.1
Experience with computer	–.11	5.5*
Computer skill level	.06	1.8
Share a terminal	–.03	.5
Ratio of workers/terminal	–.03	.3

NOTE: Multiple $R = .33$; adjusted $R^2 = .09$; $df = 9, 476$
*$p < .05$; **$p < .01$; ***$p < .001$.

One possible explanation for the lack of prediction of computer characteristics, skill, and experience, and the intervening psychological mechanisms could be the range of computer use in the sample. Although 95% of the sample reported using a computer in their work, over 30% of the users worked at a computer an average of 10 hours or less per week. To investigate whether there would be a stronger relationship among the heavy computer users between computer characteristics, use, and skill, and job satisfaction, we repeated the regression analysis using only those workers who spent at least 20 hours per week at a computer. However, we found even fewer significant effects than in the regression using the full sample. Even among relatively heavy computer users, computer characteristics, skill and experience, and the psychological mechanisms were poor predictors of job satisfaction. Routineness of work still had the strongest association with job satisfaction, but it was no longer statistically significant ($F = 2.5$, $df = 1, 103$). This result occurred because routineness was related to hours per week of computer use; when we limited our analysis to those who worked at least 20 hours per week at a computer, routineness was also truncated. People who worked at least 20 hours a week at the computer—and especially those who did so more than 30 hours a week—had more routine jobs than those who worked fewer hours a week at the computer ($F = 4.6$; $df = 3, 564$; $p < .001$).

The basic conclusion of the regression analyses was that variables of computer use, skills, and experience generally were not related to job satisfaction, and the intervening psychological mechanisms we studied were only weakly related to job satisfaction. In the case of perceived control over the computer, its weak relationship operated primarily through routineness of the job—that is, people whose job duties were repetitious reported slightly less control over the computer and were substantially less satisfied with their jobs than other white-collar workers. Symbiotic interaction with the computer, as predicted, was positively related to job satisfaction, but the relationship was a weak one.

Conclusions

In general, the workers in the 89 computer-using groups we studied were fairly recent users of computer systems. The average work group first started using computers in 1984, and the modal worker had had between one and five years of experience using computers. In addition, the computer literacy of the average worker was relatively low; he or she had only one of the six kinds of computer-related skills we assessed. For most workers, that skill was either data entry or the use of an applications package, such as word processing, spreadsheet, graphics, or statistics. The least skilled workers used computers only to read information and could not even add, delete, or manipulate information.

Over half of the supervisors reported that they usually knew how many hours a day workers spent at the computer and how many computer applications were used by their employees. Employees were a little less likely to report that their supervisor knew such information. The fear that workers will be subject to minute surveillance of their work by computers has not materialized in the present sample, although a nontrivial minority of supervisors (14 of 88)—mostly supervisors of clerical workers—claimed to have access to such detailed performance information.

Particular aspects of computer use and skills were only weakly related to the two intervening psychological mechanisms we studied: symbiotic interaction and perceived control over the computer. In general, people who were more skilled reported that they and the computer worked together *less* well than users who were less skilled, although years of computer experience were positively associated with a symbiotic interaction. The most notable finding about symbiotic interaction was the very large percentage of end-users who claimed that they and their computer system worked very well together. Over one-fifth (21.4%) of the users checked the most positive response on all five symbiotic interaction items. It seems possible that when—or if—users gain greater computer skills, they may find interaction with the computer system to be a more frustrating experience. Currently, many users perform only one or two simple functions.

The results for the second intervening psychological mechanism were even weaker. Use of and experience with computers were generally not related to the perception of control over the computer or to viewing the computer as a tool to be used at one's convenience. Computer skill level was weakly associated with the perception of control, with the more skilled workers perceiving slightly *less* control. The most important finding regarding perceived control, like symbiotic interaction, was the high level reported by most workers. Most end-users perceived that they had a great deal of control over their computer systems.

Overall, the availability of computer systems and workers' experience and skill with computers were not related to their job satisfaction. One of the psychological mechanisms, symbiotic interaction, was related to job satisfaction. The more that end-users believed they and their computer worked well together, the more satisfied they were with their jobs. Perceived control, the other intervening psychological mechanism, was not related to job satisfaction when routineness of the job was controlled. However, perceived control was associated with less routine jobs, and less routine jobs were the strongest factor associated with greater job satisfaction.

In view of the widespread belief that information technology is revolutionizing U.S. offices, it was somewhat surprising that the objective features of computer systems, such as ratio of workers to terminals or whether or not a worker has to share a terminal with others, were not related to job satisfaction. In an earlier study of computerized offices (Bikson et al., 1987), white-collar workers often reported that there was a shortage of computer equipment. When asked how their computer systems could be improved, they frequently responded with "more equipment." Similarly in the present study, many managers felt they needed more equipment, though the availability of equipment per se did not affect job satisfaction. These findings support the assumptions noted above: Computer systems are adaptable to many uses, functions, styles of management, and organizational structures. These results are consistent with the view that the computer in use is psychologically more meaningful than are specific computer characteristics. A particularly important variable is the routineness of the job. The use of computers in routine jobs was associated with lower job satisfaction; conversely, the use of computers in nonroutine jobs was associated with greater job satisfaction. Likewise, perceived control over the computer was related to routineness of the job, with end-users in routine jobs reporting less control over the computer. These workers were also likely to report that it was their job to use the computer rather than that the computer was a tool to be used at their convenience.

In sum, computers per se appear to have little direct effect on job satisfaction. The perception of control over the computer and the perception that the user and computer work well together likewise have little direct effect on job satisfaction. The computer-in-use reflects the climate of the work group. It is not the computer hardware and software—nor even the amount of the users' experience with it or perception of it—that count; how it is used is what is important. Thus computer systems mirror the existing work environment rather than alter it.

Note

1. This has not always been true. Early computers were difficult to use, temperamental, and needed an environment in which the temperature and humidity were carefully controlled. Only highly skilled operators who knew the whims of a computer were able to keep it running for more than a few hours at a time. Now systems are much more reliable and require less skill to run them. Today the computer operators of mainframe systems are classified by the U.S. Census Bureau as clerical workers.

References

Attewell, P. (1987). Big brother and the sweatshop: Computer surveillance in the automated office. *Sociological Theory*, *5*, 87-99.

Bikson, T. K. (1987). Understanding the implementation of office technology. In R. Kraut (Ed.), *Technology and the transformation of white collar work* (pp. 155-176). Hillsdale, NJ: Erlbaum.

Bikson, T. K., & Gutek, B. A. (1983). *Advanced office systems: An empirical look at utilization and satisfaction* (N-1970/NSF). Santa Monica, CA: RAND Corporation.

Bikson, T. K., Gutek, B. A., & Mankin, D. A. (1987). *Implementing computerized procedures in office settings* (R-3077-NSF/IRIS). Santa Monica, CA: RAND Corporation.

Braverman, H. (1974). *Labor and monopoly capital: The degradation of work in the twentieth century*. New York: Monthly Review Press.

Cohen, S. (1978). Environmental load and the allocation of attention. In A. Baum, J. E. Singer, & S. Valins (Eds.), *Advances in environmental psychology. Vol. I: The urban environment* (pp. 1-29). Hillsdale, NJ: Erlbaum.

Feldberg, R. L., & Glenn, E. N. (1987). Technology and the transformation of clerical work. In R. Kraut (Ed.), *Technology and the transformation of white collar work* (pp. 77-97). Hillsdale, NJ: Erlbaum.

Griffith, T. L. (1988, August). The effects of electronic monitoring on social presence and subsequent individual performance. In P. S. Goodman (Chair), *The effects of electronic monitoring*. Symposium presented at meeting of Academy of Management, Anaheim, CA.

Gutek, B. A. (1983). Women's work in the office of the future. In J. Zimmerman (Ed.), *The technological woman: Interfacing with tomorrow* (pp. 319-328). New York: Praeger.

Gutek, B., Bikson, T. K., & Mankin, D. (1984). Individual and organizational consequences of computer-based office information technology. In S. Oskamp (Ed.), *Applied social psychology annual* (Vol. 5, pp. 231-254). Beverly Hills, CA: Sage.

Hackman, J. R., & Oldham, G. R. (1975). Development of the Job Diagnostic Survey. *Journal of Applied Psychology*, *60*, 151-170.

Heider, F. (1958). *The psychology of interpersonal relations*. New York: John Wiley.

Kelley, H. H. (1973). The process of causal attribution. *American Psychologist*, *28*, 107-128.

Kiesler, S., Siegel, J., & McGuire, T. W. (1984). Social psychological aspects of computer-mediated communication. *American Psychologist*, *39*, 1123-1134.

Kling, R., & Iacono, S. (1984). Computing as an occasion for social control. *Journal of Social Issues*, *40*(3), 77-96.

Kling, R., Iacono, S., & George, J. (1989, June). *Occupational power, workplace influence, and desktop computerization*. Paper presented at Desktop Information Technology Conference, Cornell University, Ithaca, NY.

Kraut, R. E. (1987). Social issues and white collar technology: An overview. In R. E. Kraut (Ed.), *Technology and the transformation of white collar work* (pp. 1-21). Hillsdale, NJ: Erlbaum.

Kraut, R. E., Egido, C., & Galegher, J. (in press). Patterns of contact and communication in scientific research collaboration. In J. Galegher, R. E. Kraut, & C. Egido (Eds.), *Intellectual teamwork: The social and technological bases of cooperative work*. Hillsdale, NJ: Erlbaum.

Langer, E. J., & Saegert, S. (1977). Crowding and cognitive control. *Journal of Personality and Social Psychology*, *35*, 175-182.

Majchrzak, A. (1988). *The human side of factory automation*. San Francisco: Jossey-Bass.

McPadden, J. (1988). *The rate of diffusion of the personal computer in the office environment*. Unpublished doctoral dissertation, Claremont Graduate School, CA.

Miura, I. T. (1984). *Processes contributing to individual differences in computer literacy*. Unpublished doctoral dissertation, Stanford University.

Quinn, R., & Cobb, S. (1980). What workers want: Factor analyses of important ratings of job facets. In D. Katz, R. L. Kahn, & J. S. Adams (Eds.), *The study of organizations* (pp. 275-283). San Francisco: Jossey-Bass.

Quinn, R., & Staines, G. L. (1979). *The 1977 Quality of Employment Survey*. Ann Arbor, MI: Institute for Social Research.

Rousseau, D. M. (1986, August). Choosing an appropriate level of technology in organizational research. In B. A. Gutek (Chair), *Technology: Its meaning, measurement, and impact in the age of computerized work*. Symposium presented at the meeting of the Academy of Management, Chicago.

Schulz, R. (1976). Some life and death consequences of perceived control. In J. S. Carroll & J. W. Payne (Eds.), *Cognition and social behavior* (pp. 135-155). Hillsdale, NJ: Erlbaum.

Seidel, R. J., Anderson, R. E., & Hunter, B. (Eds.). (1982). *Computer literacy*. New York: Academic Press.

Shaiken, H. (1984). *Work transformed: Automation and labor in the computer age*. Lexington, MA: Lexington, Books.

Sproull, L. S., Kiesler, S., & Zubrow, D. (1984). Encountering an alien culture. *Journal of Social Issues*, *40*(3), 31-48.

Strassman, P. (1985). *Information payoff: The transformation of work in the electronic age*. New York: Free Press.

Withey, M., Daft, R. L., & Cooper, W. H. (1983). Measures of Perrow's work unit technology: An empirical assessment and a new scale. *Academy of Management Journal*, *26*, 45-63.

Working Women Institute, National Association of Office Workers. (1980). *Race against time: Automation of the office*. Cleveland, OH: Author.

Zuboff, S. (1988). *In the age of the smart machine: The future of work and power*. New York: Basic Books.

5

Informal Communication in Organizations: Form, Function, and Technology

ROBERT E. KRAUT
ROBERT S. FISH
ROBERT W. ROOT
BARBARA L. CHALFONTE

Most of the work that people do in organizations requires some degree of active cooperation and communication with others. This is true of routine clerical work and it is equally true of creative work, such as scientific research or engineering development. Indeed, in some scientific fields, over 65% of publications are jointly authored (Over, 1982), and most research projects, regardless of authorship, require support staffs of clerks, research assistants, or technicians.

Individual members of groups need to communicate with one another to accomplish their production and social functions; within organizations, groups need to communicate with other groups. The communication used is both formal and informal. Our goal as authors of this chapter is to understand the communication processes underlying group work in order to improve the communication technologies that groups have available to them. Our assumption is that by understanding how groups and organizations work and by comparing their communication needs to current communication technologies, we will be able to identify gaps in the array of communication tools that are available to people in organizations. We are especially interested in communication tools to support distributed groups. For instance, what would it take to have a nationwide task force meet and write a report as easily as if it were housed in a single building?

When we look around our places of work, we notice that informal communication seems to be a dominant activity. People read at their desks but are interrupted by phone calls. They leave to attend a department meeting but stop on the way to discuss a matter with a colleague. To answer questions about office procedure, they phone the person at the next desk rather than consult the appropriate manual. The conversations seem fluid and undesigned and yet, clearly, work is being accomplished. In looking at the contrast between formal and informal communication, it occurred to us that the more spontaneous and informal the communication, the less well it was supported by communication technology. We realized that we had well-established procedures for scheduling meetings and writing reports but little technology to support bumping into a colleague in the hall. Thus our interest was drawn toward understanding more about the nature and value of informal and spontaneous communicative activity and whether technology could be fruitfully employed to aid it.

While our attempts to understand informal communication have taken a number of empirical approaches, our interest in enabling technologies has been focused on the uses of audio-video communications media. The history of video as a communication technology has been a mixed one, showing great successes as a method of broadcasting entertainment, a mixed record as a method of disseminating educational information, and a dismal record as a mechanism for interpersonal communication. The lack of market success for such items as video telephones and video conferencing systems seems to contradict our intuitions about the value of visual contact in interpersonal communication. However, because these technologies had been primarily geared toward relatively *formal* communication occasions, we began to explore whether video's employment in systems for *informal* communication might be more successful. In particular, we thought that because video simultaneously reminds a person of a need to talk to someone and provides a communication channel through which to carry on the conversation, it might become the technology to support spontaneous, informal communication.

The remainder of this chapter elaborates our thesis that informal communication is an important mechanism to help achieve both the production goals and the social goals of groups. The chapter starts by more fully describing what we mean by informal communication, conceptually and through example, and then details some features common to episodes of informal communication. Next, the chapter examines some of the ways in which informal communication supplements more formal communication processes to aid both the production and social components of group work. While the examples and data in this chapter come from studies of informal communication in research and development (R&D) environments, we do not think the insights gained are limited to these environments. Finally, the chapter describes two experimental telecommunication systems aimed at supporting informal communication at a distance.

The Nature of Informal Communication

Theorists have long recognized that organizations have available to them communication methods varying in formality, that they tend to deploy these different methods for tasks varying in uncertainty, and that matching the informality of the methods with the uncertainty of the task leads to better organizational outcomes. At both the organizational and small-group level, the coordination of activity is the production-oriented task that has been examined in most detail. Coordination is the activity of directing individuals' efforts toward achieving common and explicitly recognized goals (Blau & Scott, 1962). As Van de Ven, Delbecq, and Koenig (1976) describe it, "coordination means integrating or linking together different parts of an organization to accomplish a collective set of tasks" (p. 322). Explicit coordination is necessary in part because individuals within an organization have only partially overlapping goals. Thus, one of the aims of coordination is to ensure that the disparate individuals come to share the same goals. But even if this aim were achieved and their goals were identical, the input-output dependencies among individuals require that their efforts be sequenced and interrelated efficiently.

The coordination mechanisms used by organizations differ in their degree of formality—that is, in their degree of prespecification, conventionality, and rule boundedness. At the formal end of the dimension, coordination is accomplished by adherence to common rules, regulations, and standard operating procedures; through preestablished plans, schedules, and forecasts; and through memos, management information reports, and other standardized communications. These formal coordination mechanisms have in common communication that is specified in advance, unidirectional, and relatively impoverished.

Informal communication is a loosely defined concept and is often treated as the residual category in organizational theory. According to this perspective, informal communication is that which remains when rules and hierarchies are eliminated as ways of coordinating activities. More positively, informal communication is communication that is spontaneous, interactive, and rich. Coordination by feedback (March & Simon, 1958), through organismic communication networks (Tushman & Nadler, 1978), or by clan mechanisms (Ouchi, 1980) are alternate ways of describing coordination by informal communication. The essence of these informal communication systems is their lack of prespecification. Information is not prepackaged and then shipped intact to a recipient; courses of action are not precomputed and then executed without modification. Rather, information is often exchanged interactively, through meetings and conversations, and courses of action are worked out in the context of the circumstances into which the actions must fit.

Figure 5.1 illustrates several of the variables that we think distinguish formal from informal communication. At the heart of what we term informal communication is its spontaneous and unplanned nature. Conversations take place at the time, with the participants, and about the topics at hand. None of these characteristics—timing, participants, or agenda—is scheduled in advance. Moreover, during its course, the communication changes to take into account the participants' current interests and understandings. In this sense, informal communication is truly interactive, with all participants in the communication being able to respond to what they perceive to be the current state of affairs, including the communication up until that point and their perception of the other participants' reactions to it. Through this feedback mechanism, informal communication can be more effective than formal chan-

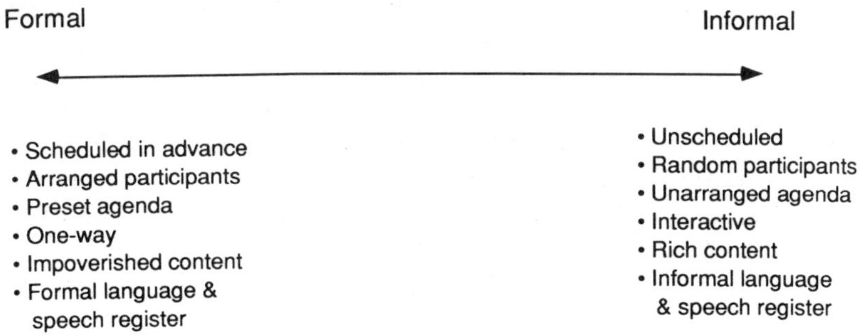

Figure 5.1. The Formality Dimension of Communication

nels because participants in the conversations elaborate or modify what they have to say in order to deal with someone else's objections or misunderstandings (e.g., Kraut, Lewis, & Swezey, 1982).

Our distinction between formal and informal communication parallels Daft and Lengel's (1984) distinction between impoverished and rich communication channels, although we add the criterion of spontaneity to their criteria of bandwidth and underactivity. According to Daft and Lengel (1986), rich communication channels are ones that "can overcome different frames of reference or clarify ambiguous issues to change understanding in a timely manner" (p. 560). In order of decreasing richness, they consider face-to-face communication, telephone, personal documents (such as letters), impersonal documents, and numeric documents. (See also Stohl & Redding, 1987, for further discussion of the formal-informal distinction, regarding communication.)

There are both structural and functional characteristics of communication occasions that cause the communication to be more or less formal. Among structural characteristics, the nature of the relationship among the participants influences its formality. For example, conversations among strangers or those with highly unequal status will be more formal than conversations among close friends or peers. The frequency of communication also influences its formality. If communication partners have the ability to communicate with each other many times a day, they need not stand on ceremony in their communication; the communication then moves from a formal to an informal style (Brown & Fraser, 1979). The nature of the communication setting also influences the formality of communication within it. A discussion in a board room is likely to be more formal than one in the corporate fitness center. Finally, the communication channel itself may partially determine the formality of a communication event. By their nature, for example, telephone and face-to-face discussions are more interactive and rich than are computer mail systems and, as a consequence, they are more informal. Subdividing media more finely, computer-generated information systems reports and human-generated memoranda are more formal than are scheduled meetings and electronic bulletin boards, which in turn are more formal than telephone calls or hallway chats.

In terms of functional characteristics, formal and informal communication systems seem best suited to different types of activities. Formal communication tends to be used for coordinating relatively routine transactions within groups and organizations. For example, in a large corporation, one might go through a procurement process simply by following the steps specified in the corporate purchasing guide. The material specification, purchase requisition forms, bidding procedures, criteria for selecting one vendor over another, and stages in the approval process would all be specified in advance. In the extreme, the rule book could so totally describe the conditions under which certain actions should occur and the precise ways of executing them that a factory's computerized, just-in-time procurement system could place orders with suppliers without human intervention.

However, these formal coordination mechanisms often fail in the face of novel or unplanned events. Novelty, unexpectedness, and uncertainty are frequent in organizations and are often components of what appear to be routine procedures (e.g., Suchman & Wynn, 1984). Under these circumstances, informal communication seems needed for coordination in the face of uncertainty and equivocality (Daft & Lengel, 1986). Thus, while a group might purchase a new desk by delegating the responsibility to one member and then following the purchasing rules, they would not think of hiring a division manager simply by following the procedures laid out in the personnel guide. Instead they would have plenty of informal communication with one another about the type of person they wanted as a leader and with the candidates to assess their qualities. Informal communication is needed in this case because, compared to buying furniture, personnel procurement is a rare event, with considerable consequences, and it is also highly uncertain, given the difficulty of predicting the characteristics of people.

Research findings lend support to these hypotheses. For example, when work groups are engaged in more complex tasks—that is, tasks that are varied, lack routine procedures, and require group members to think through solutions—such groups are far more likely to communicate directly with other group members and to have more scheduled and unscheduled meetings to coordinate their activities (Van de Ven et al., 1976). Daft and Lengel (1984) have shown that organizational members prefer rich and interactive media, such as face-to-face meetings, when they have value conflicts and other disagreements to work out. And Argote (1982) has shown that when groups with greater task uncertainty do engage in meetings and unscheduled communication to coordinate their activity, they are more successful in performing their work than if they rely on standard procedures.

In sum, we have argued that informal communication supports organizational and group coordination, especially under conditions of uncertainty. Coordination is an example of a production function of groups. When people work in groups within organizations, they must achieve three goals to be successful: production, group maintenance, and member support (Hackman, 1987; McGrath, 1984, in press). They must actually accomplish productive work—write the reports, make the decisions, construct the software, allocate the budgets, defend clients, or do whatever the particular group is assigned to do. In addition, groups must achieve two social goals. First, they must sustain themselves over time, and most groups have life spans of multiple projects. To sustain themselves, groups have to recruit and socialize members, keep them happy enough so that they want to maintain membership, garner external resources, and do the sundry other activities that ensure the group's continuing survival. The second social goal that the group must attempt to achieve is to support the needs of individual group members so that they feel satisfied with their work, relationships, and membership in the group. By most criteria, a team whose members are unhappy with their work and hostile toward one another would be deemed a failure, even if they accomplished their tasks. In scientific research teams, for example, scientists must feel that they are making useful contributions and that their contributions are being recognized; such teams often break up if this recognition is not forthcoming (Kraut, Galegher, & Egido, 1988).

There is reason to think that informal communication is particularly useful in supporting the social functions of groups. This is because organizations are less explicit in regulating social relationships than in regulating other aspects of work procedures. For example, corporate personnel guides frequently describe the bureaucratic procedures for annual performance appraisals, but they neither attempt to nor could they regulate the ad hoc personal judgments that supervisors make of the people reporting to them. A vast literature in social psychology suggests that relatively unstructured and informal communication is at the basis of the social processes, such as person perception and liking, which underlie group maintenance and member support (e.g., Festinger, Schacter, & Back, 1950; Zajonc, 1968). To give one example, Gabarro (1987) describes the development of mutual expectations and trust that sustains the work relationships among managers; when a CEO no longer trusts the judgments of his or her subordinates, the subordinate is frequently transferred, demoted, or fired. "[Mutual expectations] . . . are . . . typically worked out over time during a succession of routine interactions, such as ad hoc encounters, meetings, progress reviews, and discussions of task-based problems" (p. 184). These are the mechanisms of informal communication.

The Dynamics of Informal Communication in R&D

Moving from abstract discussion of communication styles, groups, and organizations in general, this section, as well as subsequent ones, will discuss several examples of informal communication and examine its functions in the context of the research and development process among scientists and engineers. We focus on informal communication in research and development for two reasons. First, as we have seen previously, most research is collaborative, requiring coordination and communication. Second, the central feature of this domain is dealing with uncertainty in both production and social relations. In terms of production, the object of research in both science and engineering is to create novelty. In embarking on a research project, scientists and engineers must make the equivocal judgments of whether their goal is a valuable one and whether their methods are appropriate and will meet with success. Even when the project is completed, they are still left with these issues, as suggested by the recent controversy concerning the value of research on room-temperature nuclear fusion. In terms of social relations, the uncertainties are in establishing trust in potential and actual research collaborators and in defining an equitable division of labor and credit for jointly planned work (see Kraut, Galegher, & Egido, 1988, for a more complete discussion of these dilemmas in scientific research). For these two reasons—the importance of collaboration and the essentially uncertain nature of the research enterprise—we believed that informal communication would play an extremely important role in research and development; as a result, the R&D domain would provide an interesting lens through which to view the dynamics and functions of informal communication.

In this section we describe three examples of informal communication. The episodes come from samples of videotaped interactions we collected from hallways, copy centers, entryways, and other commons areas in one building of a large R&D laboratory. These episodes were collected by turning on an unattended videotape camera at random times for hour-long intervals. The camera was large, plainly visible, and identified with a sign stating that recording for research purposes was occurring. During most of the 12 hours that were recorded, there was no social interaction at the recording sites because either no one was present or only one person was. When two people were present, most social interactions were minimal. A typical scenario would be for two people to walk past each other in the hall, acknowledging each others' existence by a stereotyped pattern of glances toward and away from each other (Goffman, 1963) or with a brief "hello" in passing.

Given this context, the excerpts presented below are by no means random selections from episodes in which at least two people were videotaped. Rather, they were selected because they illuminate both the dynamics and functions of informal communication. Thus, they provide a useful counterpoint to the more schematic and quantitative descriptions of informal communication presented in later sections.

Episode 1: The Call

The first episode illustrates the beginning of what we have called an opportunistic, work-related conversation. Andrew is returning to his office when he sees Bob backing out of a meeting in an office at the end of the hallway. Bob is unaware of Andrew's presence. Andrew calls to Bob, timing his call by observing when Bob has concluded his participation in the meeting. Bob turns around, recognizes Andrew, and acknowledges his presence with a smile. He starts walking toward Andrew, who remains stationary until Bob reaches him. In the final scene of the episode, Andrew and Bob are walking off together down the hallway, discussing the difficulties they have had contacting a mutual colleague.

Both Bob and Andrew abandoned their original goals (concluding a meeting or returning to an office, respectively). As a result of serendipitously seeing each other in the hallway, they were able to pass information about the status of a project and to solve a problem that was hindering the project's progress. Andrew told Bob that he had not been able to contact a colleague, and Bob suggested a way of doing so. This is a common mechanism through which small collaborative teams do project management (see Kraut, Galegher, & Egido, 1988, for a description of communication in scientific research teams). For small groups, this informal project management serves the production function of coordination and problem solving efficiently, assuming team members run into each other enough. It has the additional benefit of keeping group members informed of and involved with many of the minor decisions and crises that occur in any project. Thus it serves the social functions of groups, by keeping group members committed to the projects of which they are a part.

Episode 2: Stop Short

Like the previous episode, Episode 2 shows the chance initiation of a work-related conversation. In addition, it also demonstrates how people pick up background information about their work environment by merely navigating its hallways. A typical sequence observed as people walk down the halls is for them to peer into open offices and public spaces as they go to the printer, copy machine, bathroom, or other destination. They usually do not slow down, but simply turn their heads as they pass open doorways. This process of browsing the social environment while on other business provides people with a substantial amount of information about the world in which they live. They learn or are reminded of who the local inhabitants are and the goings-on in their offices give some indication of their characteristics, interests, and current activities. While much of this information is not immediately relevant to the tasks at hand, these observations of the workplace provide some of the background knowledge through which people make sense of the subsequent information they acquire. For example, they provide a basis for the mutual knowledge that people need to understand one another (Clark & Marshall, 1981; Krauss & Fussel, in press) and the firm-specific knowledge that employees acquire in the first years of working for the company.

In this episode Abel, a researcher, is walking down the hallway, glancing into offices as he goes. He sees another researcher, Baker, whom he knows, notices that Baker has spotted him, and says hello while passing by. The next scene occurs seven seconds later when, even though Abel has already passed Baker's office, two phenomena cause him to reverse his steps and stop to have a brief discussion with Baker. First, the mere sight of Baker served as a stimulus that jogged Abel's memory that he had something to say to him. Second, the "hello" served as a channel checking routine, indicating that Baker's attention was free and that he was available for conversation. This confluence of topic and availability, which Baker's visual presence provided, was the minimum precondition for a conversation to occur.

In the third scene, another researcher, Charlie, walks down the hall, also glancing in offices as he goes. He observes Abel and Baker in conversation, and turns his head farther to see them as he is walking. Although he does not slow down, his actions indicate that this is an interesting event for him. This observation provides Charlie with background information that may be useful in the future. By catching a glimpse of the participants in ongoing conversation, Charlie may be able to make inferences about the relationship between the participants or the topic of the conversation. In this way he learns about social relationships, alliances, crises, and collaborations in the laboratory. In the final seen, Abel finishes his discussion with Baker and continues on to his original destination. During this 35-second sequence, Abel and Baker, who had not planned to speak, got a little bit of work done and provided some information to Charlie, which he can use to better understand his work environment.

Episode 3: At the Candy Machine

Episode 3 is more blatantly social than the previous episodes. In the first scene, Ann and Barry are having an informal, social discussion in front of some vending machines as they take a break from work. Barry sees a coworker, Chester, walking by on his way to the elevator and hails him. Chester stops, and Barry, with Ann in tow, approaches him. Ann, who did not know Chester, is introduced, and the three have a discussion regarding Chester's wedding. Barry asks, "When's the big day?" He recognizes that Ann does not know Chester and explains that Chester is to be married. Chester replies that his wedding was last weekend, and both Ann and Barry congratulate Chester.

Three functions are served in this brief episode. First, Barry is maintaining social bonds with two coworkers, Chester and Ann. These minor social pleasantries are undoubtedly useful in developing and maintaining social networks in organizations. They form the basis of many mutual obligations and work exchanges in organizations. For example, if Barry needs information, it should be easier for him to call upon either Ann or Chester for help than it would be if they had never engaged in these social pleasantries. Second, Barry acted as a social catalyst by introducing Ann and Chester and providing information to each about the other. Ann should now have an easier time talking to and dealing with Chester than she would have if the introduction had never been made. Finally, during the discussion of Chester's wedding, Barry inadvertently reveals to Ann information about the closeness of Barry and Chester's relationship: They are coworkers who know each other, but not well, and who marginally keep up with each other's affairs through work place gossip. Ann can store these facts as potentially interesting background information that helps her understand her work environment.

Shared Features of the Episodes

None of these brief episodes of informal communication was especially significant in its own right. If any one of them had failed to occur, no careers would have been ruined or projects stymied. Yet episodes like these occur hundreds of times a day in both small and large departments. We believe that in the aggregate, they are fundamental methods that organizations use to get their work done, transmit organizational culture and firm-specific knowledge, and maintain the loyalty and goodwill of their members.

The episodes reviewed here share several noteworthy features. First, all were unplanned and unanticipated. The participants all engaged in more or less useful conversations, but they did not know such conversations would take place even seconds before they occurred.

Second, the visual channel was a prerequisite for these interactions to occur. In each example, the visual channel was instrumental in identifying a partner for conversation, in identifying the precise moment when the potential partner was available, and in establishing a topic for the conversations. In each case, the initiator of the conversation first saw and recognized a potential partner whom he already knew. As Kendon and Ferber (1973) describe it, sighting is a necessary preinteraction phase of a greeting. Recognizing someone to talk to is not simply a passive process of following one's mother's advice to refrain from talking to strangers. Rather, seeing someone often brings with it a social obligation to acknowledge that person's presence with a greeting. In the candy machine example, Barry seemed to be responding to this social obligation as he went out of his way to greet Chester.

In addition, the visual channel was used to identify an opportunity for conversation. By looking at a potential target of conversation, the initiator can often interpret the target's locus of attention and infer whether and when he or she is available for conversation. In the call example, the initiator could see when the target had disengaged from his previous conversation and was available to be hailed. Potential conversationalists often use the visual channel to synchronize their behaviors (see Kendon & Ferber, 1973) to make sure there is a clear channel between them. For example, in the call episode, after Andrew hailed Bob, he did not engage in conversation until Bob had turned around, engaged in eye contact, and acknowledged his presence with a smile.

Seeing someone also serves as a potent stimulus to evoke topics of conversation. In the call episode, the sight of Bob appeared to remind Andrew of project information of which Bob should be aware. In the stop short episode, seeing Baker and saying hello to him clearly reminded Abel of a topic of conversation, and he reversed his tracks to follow it up. In the candy machine example, Barry's sighting of Chester appeared to remind him of Chester's wedding and Barry's social obligation to inquire about it. Thus the visual channel is often instrumental in establishing topics of conversation.

A third characteristic illustrated by the examples is that both production and social work often get done in these informal conversations. Though one might characterize a conversation as having a main purpose, the production and social goals frequently co-occur; whether a conversation is work-related or social is a matter of degree. Sometimes a conversation is directly related to production. In the call episode, the participants coordinated their activity, informed each other about the status of their project, and solved a minor problem that was impeding progress. Yet it is clear from the smiles and positive effect displayed in this episode that the participants were enjoying each other's company as well. Other times, a conversation is basically social, as in the candy machine example, where members of an organization were able to establish and reaffirm their bonds to each other. But even here, it appears that coworkers were forging bonds that would underlie later work activity. Many times, informal conversations do not directly support the work at hand. However, even these often provide background information that might be useful for completing a work assignment or acquiring more general organizational competence (Granovetter, 1973).

Surface Characteristics of Informal Conversations

The following two sections describe some features and functions of informal communication more systematically and quantitatively. Casual observation suggests that brief encounters of the sort described above are extremely common in organizations. We concentrate (although not exclusively) on these brief and unplanned encounters because they are so common. Like an ethnologist, we start with the assumption that if a behavior pattern occurs frequently enough, it is likely to be important for a species or group. As social observers, we have a responsibility to describe these behavior patterns and to understand functions that they might serve.

Some features of these conversations are easily observable. These surface characteristics, as we will call them, comprise information about informal communication that can be obtained through simple methods. In this section we present descriptive data about the surface characteristics of informal communication—specifically, how frequently informal conversations occur, how they are initiated, where they are likely to take place, and how long they last.

If we use the frequency with which informal communication occurs as a measure of importance, then numerous studies have shown that informal communication is vitally important in organizations. Since at least the time of Mintzberg's (1973) study, organizational researchers have observed that informal communication is the dominant activity of managers. Sproull (1984) reviewed the evidence from seven studies of managerial communication in which managers ranged from mid-level rank (e.g., factory heads and school principals) to the most senior (e.g., CEOs of moderate-sized corporations and college presidents). Data collection included both direct observations of the managers by researchers who shadowed them during their workday and managers' self-reports of communication in daily diaries. Across the seven studies, verbal interaction accounted for about 75% of the managers' workdays. About 50% of this verbal interaction consisted of unscheduled face-to-face meetings with an additional 12% consisting of unscheduled telephone calls. Together these figures indicate that almost 50% of the typical manager's time is consumed by the unscheduled conversation that we consider informal communication. In general, informal communication appears to be a frequent and therefore important activity through which managers find out information, communicate opinions, and make decisions.

This characteristic of managers is also true of researchers in R&D environments. We make this statement based on data we collected in a study of conversations occurring in buildings of an industrial research laboratory and a state university. The aim of the study was to examine the characteristics of a sample of face-to-face interactions among members of these two organizations. We identified conversations occurring in a sample of locations in buildings within the organizations at randomly determined times. When a conversation was identified by a researcher, participants in the conversation were asked to complete a brief questionnaire describing it.

To sample conversations, a researcher first identified blocks of seating locations in a building, such as offices located along a corridor or desks within a large, group office.[1] The researcher went to these blocks at random times and made a single pass through the area (i.e., each individual was observed only once). The researcher interrupted every face-to-face conversation observed during the sampling period, explaining, "We are trying to understand what functions conversations serve in organizations," and asked all people involved in the conversation to complete a brief questionnaire at the end of the conversation. The questionnaire requested a description of the conversation (e.g., how it came about, where it occurred, its length, and its topic), the usefulness of the conversations for various functions (e.g., maintaining a working relationship, coordinating work, keeping up with workplace information), and the relationship between the participants (e.g., their organizational ties, project status, and frequency of other communication).

The researchers identified 522 sites (occupied desks or offices) where conversations could have occurred and observed 695 individuals. At the time of observations, 121 (23%) of these sites actually contained a conversation representing 267 individuals (38%). We had at least one usable response from 117 of the 121 conversations (97% response rate). In terms of individuals, the response rate was 83%.

Frequency and Initiation of Informal Conversations

At any moment in the R&D organizations we observed, 38% of the people present were engaged in face-to-face conversation. As one index of the informality of conversation, we asked respondents to indicate how these conversations came to be. This was operationalized as the degree to which the conversation was scheduled or spontaneous. Unlike previous researchers (e.g., Van de Ven et al., 1976), we distinguished among degrees of spontaneity, assuming that the more spontaneous the conversation, the more informal it was likely to be on many dimensions. All participants were asked to categorize the degree of preplanning that characterized the conversation in which they had just been involved. The four categories were: (a) a conversation that was previously scheduled or arranged (we term this *scheduled*), (b) one in which the initiator set out specifically to visit another party (*intended*), (c) one in which the initiator had planned to talk with other participants at some time and took advantage of a chance encounter to have the conversation (*opportunistic*), or (d) a spontaneous interaction in which the initiator had not planned to talk with other participants (*spontaneous*). Each respondent characterized the meeting individually. Because each participant was likely to have a different view of how the conversation began, we organized the data by conversation and assigned an initiation value to the conversational unit. This value was defined as the most preplanned (least spontaneous) of any respondent's classification (ordered scheduled < intended < opportunistic < spontaneous on the spontaneity dimension). Thus, if the initiator of a conversation categorized it as intended and the other party classified it as spontaneous, we categorized the conversation as a whole as intended. This scheme provided a conservative estimate of spontaneity and also handled rating differences based on the role of the participants in the conversation.

The distribution of 117 conversations that we identified was as follows: 12%, scheduled; 36%, intended; 21%, opportunistic; and 31%, spontaneous. If we assume that only scheduled meetings are truly formal, then by this definition 88% of the conversations we sampled were informal. This is a conservative estimate because many scheduled meetings, particularly in an R&D environment, contain elements characteristic of informal communication. For example, a research planning meeting is typically highly interactive. This type of scheduled meeting is in sharp contrast with the elaborately choreographed activity at an annual shareholders meeting or a product-rollout meeting.

The 36% of conversations that were classified as intended by at least one of the parties are functionally equivalent to ambulatory phone calls, in that one person has intentionally gone in search of another in order to have a conversation. From the perspective of the intended object of this search, however, the conversation may appear entirely spontaneous because he or she had not anticipated its occurrence. The remainder of the conversations, about 50% of the total, happened by chance, with no preplanning on the part of any participant in the conversation. These conversations would not have occurred had the participants not wandered past each other's offices on the way to the copier machine, chanced upon each other in the hallway, or otherwise bumped into each other as they moved around in their physical environment. About 40% of chance encounters (21% of the total) were opportunistic; that is, they were triggered by the sight of a person with whom one party had previously formed an unacted-upon intention to talk. In the remaining 31% of the cases, no party to the conversation had anticipated its occurrence, and it was only the opportunity for conversation presented by the physical proximity of the participants that led to the spontaneous initiation of conversation.

Initiation of conversation seems to be a joint function of the salience or importance of the conversational topic and the ease of the conversational execution. In the case of both scheduled and intended conversations, the topic is important enough for somebody to initiate action that will lead to a discussion of that topic; groups schedule a meeting or individuals leave their offices in search of a suitable conversational partner. Opportunistic conversations are more complex. In some cases, there may be no topic that is important enough to warrant immediate action, and a conversation is simply deferred. Often, however, a manager or researcher in the course of making a decision feels the need to consult with a colleague, but other events or thoughts intrude (see Mintzberg, 1973; Reder & Schwab, 1988; Sproull, 1984, for evidence on the short attention spans of managers and researchers), and both the decision and the need to consult are placed on hold. In either case, when the decision-maker later comes across the colleagues in the hallway, two phenomena occur. Upon seeing the colleague, simple associative mechanisms remind the decision-maker of the original need for consultation and increase the temporary salience of the suspended decision, perhaps even reinstating the original decision-making context. Second, being in the colleague's presence simultaneously lowers the cost of communication. The decision-maker can see whether the colleague is available and has a clear channel through which the conversation can start (see Kendon & Ferber, 1973, on the initiation of interaction).

Spontaneous interactions are even more interesting. In these cases, there is no previous topic or need for consultation. Instead, the mere opportunity for conversation created by the presence of a suitable partner and availability of a clear channel serve to generate a conversation. Observations that we will review later in this chapter suggest that conversations, especially spontaneous ones, typically occur among people who already know each other from other contexts. At any occasion, they may have nothing in particular to say to each other, but norms of politeness require that they acknowledge each other's presence and perhaps exchange greetings. These passing social encounters occasionally evolve into substantive conversations. The topics for these conversations may be based on other activities in the environment or some other experience of joint interest that is remembered or generated during the course of the conversation (the candy machine conversation described previously is a typical case).

Location of Informal Conversations

Conversations that differ in their degree of spontaneity also differ on other characteristics. First, most conversations involve people who are housed close to each other. In our survey, each participant reported the location of the sampled conversation in relation to the location of his or her office. The choices given were: (1) the participant's office, (2) next door, (3) the same corridor, (4) the same floor, (5) the same wing/section of building, (6) a different wing/section of building, (7) the same campus/site, (8) a different campus/site. Each conversation was coded to reflect the farthest distance that any participant in the conversation had to travel in order to participate in that conversation.

Workplace conversations are, in general, quite local events, usually involving people who are in close physical proximity. As Figure 5.2 shows, 52% of all conversations involved people located within the same corridor, and 87% of them took place among people who shared the same floor in a building.

Our data are thus consistent with other observations showing the exponential decay of conversations with physical distance. Indeed, Zipf's (1949) principle of least effort predicts that the closest colleagues will have the most conversations. Whether it is because random motion causes people to have greater interaction by chance with those to whom they are closest (because people intentionally try to limit their contact efforts to those who are closest), or because organizations co-locate people who have the most need for communication, studies have shown time and again that physical proximity is strongly associated with frequency of interaction (see Allen, 1977; Festinger, Schacter, & Back, 1950; Kraut, Egido, & Galegher, in press; Zipf, 1949). Floor boundaries are an especially important functional barrier for all types of face-to-face communication.

These distance effects, however, are more powerful for more spontaneous conversations. In general, the data show that the more spontaneous the conversation, the greater the likelihood that participants' offices were located close to each other ($r[111] = .23$, $p < .02$). For example, fully 40% of the spontaneous conversations occurred among inhabitants of the same office. Of the spontaneous and opportunistic conversations, 91% occurred among people on the same floor, while only 82% of the combined intended and scheduled meetings did so.

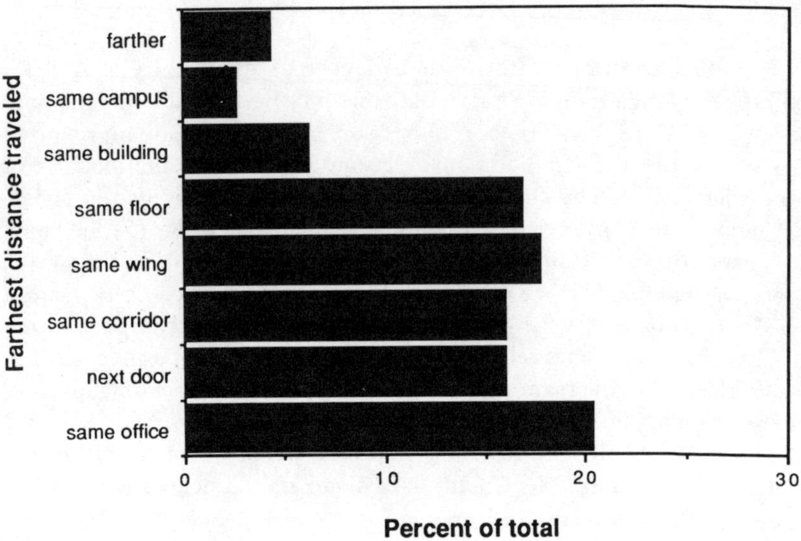

Figure 5.2. The Location of Conversations (N = 113)

It is probably for this reason that the more spontaneous conversations were composed of people who frequently communicate with each other. We asked participants to indicate how frequently they had spoken to their conversational partner in the preceding five workdays; we scored the mean of their responses per conversation. In general, most of these face-to-face conversations were but one of a series. The mean frequency of communication was 12.5 times in the preceding five days ($s.d.$ = 14.6), but the more spontaneous (and therefore informal) conversations tended to occur among those who talked together most frequently ($r[112] = .18, p < .06$).

Duration of Informal Conversations

Differences in spontaneity and therefore in the degree of informality of a conversation were also associated with differences in conversation duration. Each participant in our survey estimated the length of the conversation in which they were involved; the mean of all participants'

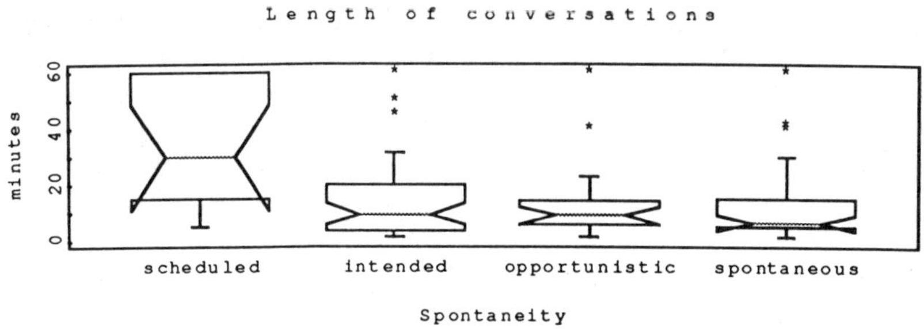

Figure 5.3. Duration of Conversations, Shown by Box Plots

NOTE: The waist of each box plot represents the median, and the notch around the waist represents one standard error on either side of the median. The box itself contains the middle half of the data, while the whiskers extending from the box reach to the most extreme nonoutlier. Outliers are plotted individually.

estimates was used as the duration for a given conversation. To reduce outliers in the data, conversations that lasted longer than 60 minutes were assigned a length of 60 minutes. In general, the more spontaneous the conversation, the briefer it tended to be ($r[115] = .26, p < .02$). In particular, Figure 5.3 shows that scheduled meetings were substantially longer than other types of conversations which, in turn, did not differ from one another. The median length of a scheduled meeting was about 30 minutes, while intended, opportunistic, and spontaneous meetings tended to last less than one third of this time—each with a median of less than 10 minutes.

Summary

Some of the characteristics of informal conversations are readily apparent from observations of conversations in progress. Informal communication tends to be frequent, accounting for over 85% of the interactions in our sample of conversations in R&D environments. While some of the informal interactions we sampled (about 35% of all conversations) were intended in that one of the participants sought out another, about 50% of the conversations were unplanned in that participants did not know they were going to speak to each other until they physically happened across each other. This result is consistent with the principle that informal communication, like most other sorts of communication, is distance sensitive. In other words, it happens most often with people who are in physical proximity. Finally, informal conversations tend to be shorter than more formal communications. This tendency may be indicative of the types of uses to which informal communications are put in the workplace, a topic we take up more fully in the next section.

The Content and Uses of Informal Conversation

Having looked at some of the surface characteristics of informal conversations, we now turn our attention toward understanding the uses to which such conversations are put. This includes understanding something about the content of conversations that we consider informal as well as the role that conversations of this sort play in the functions of the workplace. Four studies are represented below that investigate the functions for which people perceive informal communication to be useful, the effects informal communication has on familiarity with and liking for one's coworkers, the effects on collaboration when informal communication is not allowed, and, finally, the relationship between informal communication and the success or failure of research collaborations. As each of these topics is discussed in turn, we describe the methods of the studies that provide the relevant data.

Perceived Value of Informal Communication for Production and Social Functions

Perhaps the most direct way of determining the value of conversations of different sorts is to ask participants to evaluate this immediately after they have occurred. We employed this strategy for 57 of the conversations from the study described above in which researchers sampled ongoing conversations in an R&D environment. Using 5-point Likert scales, respondents rated the outcomes of their conversations. Scores for a conversation were based on the mean judgment of all participants in that conversation. Adopting a variant of McGrath's (in press) model for group functions, we measured the degree to which conversations supported the production and social functions of group work, including both member support and group maintenance. The production function was measured by a 5-item scale with such questions as, "How useful was this conversation in getting your work done?" "How productive was this conversation?" and "How useful was this conversation for coordinating your work?" The social function of the conversation was measured by a 3-item Likert scale with the following questions: "How enjoyable was this conversation?" "How much did you learn about each other through this conversation?" and "How useful was this conversation for maintaining a working relationship?" The production measure was quite reliable (Cronbach's alpha = .89); the social functions measure was less so (Cronbach's alpha = .46). While we expected that the theoretical distinction between production and social functions of a conversation would be sharply reflected in the rating responses, results of a factor analysis indicated otherwise. In particular, the question about maintaining a working relationship loaded on both the production and the social factors. As a result, the two scales were moderately correlated ($r = .26$), lending support to the idea that both social and production functions are intertwined in real work groups. If one is going satisfactorily, the other is apt to do so as well.

A repeated-measures analysis of variance on these data revealed that, overall, the more spontaneous a conversation, the less valuable it was perceived as being (F [1,55] = 3.9, $p = .05$). However, this conclusion must be tempered by the interaction revealed by Figure 5.4, which shows that the effect for spontaneity exists only for the production dependent variable (F[1,55] = 3.5, $p < .07$). Figure 5.4, top panel, shows that both scheduled meetings and intended meetings were perceived as more valuable for getting work done than were opportunistic meetings or spontaneous meetings ($t(52) = 2.6$; $p = .01$). The relationship between preplanning and perceived value for production was partially a function of the length of the meetings, as the data above have demonstrated. Planned meetings were longer and, as a result, allowed more time for planning, discussion, and decision making. However, the relationship between preplanned and perceived value for production remained even when the duration of the meeting was held constant through multiple regression (standardized beta = .37, df = 1,54, $p < .01$). On the other hand, as shown in Figure 5.4, bottom panel, the value of conversations for social functions was unrelated to their spontaneity (standardized

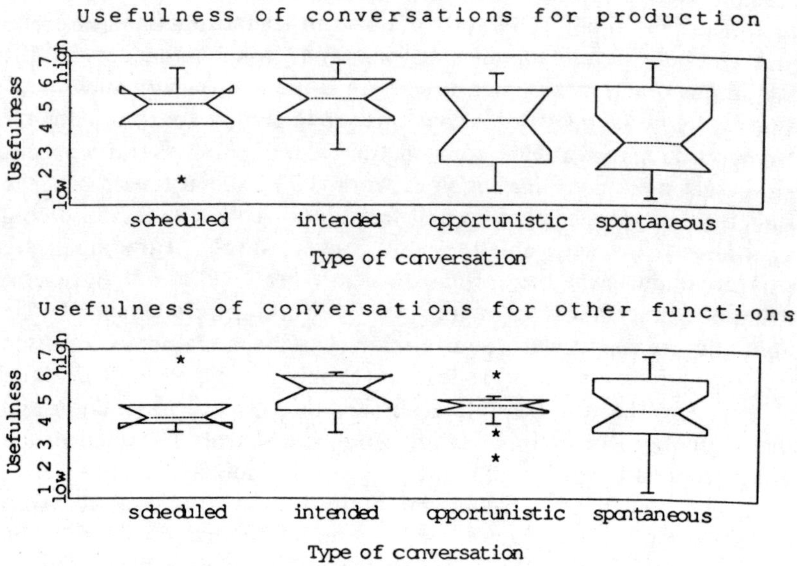

Figure 5.4. Value of Conversations for Production and for Social Functions
NOTE: For explanation of box plots, refer to Figure 5.3.

beta = .008, *df* = 1,54, $p > .50$). Brief hallway encounters were as valuable as hour-long, scheduled meetings in terms of getting to know coworkers and maintaining working relationships with them.

If one simultaneously considers the value of different types of conversations, the frequency with which they occur, and their duration, it is clear that much important work in R&D organizations occurs in unscheduled meetings. Both our own data and the earlier literature show that much of the detailed planning associated with a product gets done in relatively long, arranged meetings in which participants wrestle with their research ideas and set research directions (Kraut, Galegher, & Egido, 1988). Yet many smaller decisions and much of the coordination during the execution of the project itself get done in briefer and more spontaneous encounters. In terms of production, intended meetings are as valuable as scheduled ones for getting tasks accomplished; they occur four times as frequently as scheduled meetings, yet on a per meeting basis take only one third as much time to accomplish. Scientists, engineers, and managers use these meetings as "just-in-time" consultation and decision-making opportunities. When they are working on a problem and need information or advice, they go down the hall and solicit it during a brief chat, rather than scheduling the time, agenda, and participants in advance.

In terms of social functions, all types of conversations provided some opportunity to enjoy the company of coworkers, to learn more about them, and to build bonds with them. Again, though, because scheduled meetings occur relatively infrequently and consume greater amounts of time, they fulfill the social functions inefficiently. Here, the brief and informal intended, opportunistic, and spontaneous conversations have an edge because they represent the vast majority of conversations in organizations.

The Effects of Informal Communication on Person Perception

Much research in both industry and academia is developed through voluntary collaboration. A fundamental requirement for these collaborations to occur is for researchers seeking partners to identify others with appropriate research interests and personal characteristics (see Hagstrom, 1965; Kraut, Galegher, & Egido, 1988, for a fuller discussion of the initiation of research collaborations). All else being equal, researchers want to work with people who are competent enough to contribute effectively to a joint project, trustworthy enough to do their share of the work and take only their share of the credit, and likable in their personal characteristics. Here we test the hypothesis that informal communication is the mechanism that researchers use to size up the pool of potential collaborators before becoming committed to working with any one of them.

We have seen that communication, especially unscheduled communication, increases as the distance between conversational participants decreases, and that participants in conversations report that unscheduled communication is especially useful for supporting the social functions of groups. Our hypothesis is that increases in informal communication will also have an effect on one's perception of coworkers; in particular, the more one communicates with someone, the more one is familiar with their work, is accurate in judging them, and likes them personally. These questions were examined in a survey of 51 scientists and engineers within a single building in a research and development laboratory. The major purpose of the survey was to examine the relationship of informal communications among these researchers to measures of person perception and attitudes toward each other. For the purposes of this study, we treated physical proximity as a surrogate for frequent, informal interaction.

Individuals partaking in this study completed questionnaires in which they rated their familiarity with other participants and with the work of those participants. In addition, for those persons in the study with whom they were sufficiently familiar, they rated the degree to which they liked those individuals and the degree to which they found their work to be interesting, important, and well executed. In the analyses that follow, results are reported as mean Pearson correlations between proximity and other measures of interest. Because each participant had a spatial relationship to and judged 50 other participants, we can compute a Pearson correlation between proximity and the person perception measures for each participant. Analyses are based on the means of these individual correlations, corrected by the Fisher formula.

Indices of Proximity

Participants in the survey were located on two floors of a single research building. We developed two proximity measures for each respondent. Like Allen (1977), Festinger, Schacter, and Back (1950), and Monge and Kirste (1980), we treated proximity as more than the linear distance between offices. Rather, functional proximity is the opportunity to engage in interaction and is moderated by architectural and organizational features as well as linear distance. We assumed that locations on different corridors, sections of the buildings, or floors were separated by more than physical distance. The first index of proximity (within-floor proximity) for each pair of participants was the shortest-path distance[2] between office locations (adjusting these distances if the shortest path crossed hallways or wings of the building) × –1. Approximately half the targets of person perception had offices located on the same floor as a rater, and half had offices on an adjacent floor. We assumed that for each respondent, targets on the same floor were functionally closer than targets on an adjacent floor. Thus, the second proximity index (between-floor proximity) was coded 1 if a target was on the same floor and 0 if the target was on a different floor.

Familiarity

Using 7-point Likert scales, respondents rated how familiar they were with 50 other members of the organization and their work. In addition, if they were sufficiently familiar with these organizational members and their work, respondents had the opportunity to rate how much they liked each of these members and respected their technical work. We treated willingness to make the ratings as a sign of familiarity and failure to do so as a sign of lack of familiarity. The correlations among these four measures ranged from .70 to .84. We therefore constructed a single measure of familiarity from these measures by standardizing and averaging them (Cronbach alpha = .87).

Participants were much more likely to be familiar with colleagues on the same floor than with colleagues on an adjacent floor (the mean Pearson correlation between the familiarity scale and between-floor proximity = .41, $t[50] = 19.6$; $p < .0001$). For example, respondents thought that they knew enough about their colleagues' technical work to judge it for 42% of their colleagues whose offices were on the same floor as theirs, but made this judgment for only 17% of their colleagues on an adjacent floor. For researchers with offices on the same floor, the closer their offices, the better they knew each other; however, this proximity effect within floors was smaller than the between-floor effect (the mean Pearson correlation between the familiarity scale and within-floor proximity = .11, $t[43] = 3.8$; $p < .001$). On the other hand, among colleagues on different floors, further increases in distance had no relationship to judgments of familiarity. Essentially, when two people were already a floor apart, increasing the distance by a hallway or more had very little effect on their perceived familiarity.

Liking

Just as proximity was associated with familiarity, it was also associated with one's liking for colleagues and respect for their work. Among those whom respondents knew well enough to judge, participants reported liking colleagues on the same floor far more than their colleagues on a different floor (the mean Pearson correlation between liking and between-floor proximity = .40, $t[50] = 118.1$; $p < .0001$). Respondents tended to like their colleagues on their own floor more the closer their offices were to those of the respondents (the mean Pearson correlation between liking and within-floor proximity = .08, $t[46] = 1.87$; $p < .07$).

The results for respondents' respect for other people's work showed a similar pattern. They had greater respect for the technical work produced by people on their own floor than for work produced on a different floor (the mean Pearson correlation between the respect for work and between-floor proximity = .31, $t[50] = 13.2$; $p < .001$). Moreover, for others with offices on their own floor, respondents tended to have more respect for the work of people closer to them than for people farther away (the mean Pearson correlation between the respect for work and within-floor proximity = .09, $t[44] = 1.68$; $p < .10$).

Summary

We have shown that physical proximity, which we have treated as a proxy for frequent and informal communication, is associated with several measures of person perception. As the opportunity for informal communication with colleagues increases, so does one's familiarity with them and their work, as well as liking for them and their work. While these findings are not new (see Allen, 1977; Bossard, 1932; Maisonneuve, Palmade, & Fourment, 1952; Segal, 1974) and our use of physical proximity means they are not definitive, either (e.g., Berscheid & Walster, 1969; Pelz & Andrews, 1966; Tajfel, 1970; Zajonc, 1968), they are interesting because they suggest the power that informal communication and the physical proximity that supports it can have on important social psychological processes in organizations.

Informal Communication and Momentum in Collaboration

Another method of gaining some insight into the uses of informal communication in the workplace is to investigate situations in which informal communication channels are limited or eliminated. We explored this possibility by conducting an experiment comparing two small work groups, one of which could use both formal and informal means to communicate while the other was limited to relatively formal mechanisms. The first of the groups was successful and the second was not. We suspect that their difference in success was caused by their differences in informal communication, although a case study cannot prove this suspicion.

The experiment compared two collaborations, a "remote" collaboration that was constrained to use relatively formal communications channels, and a "standard" collaboration that used both formal and informal channels (Fish, 1988). Both collaborations involved researchers interested in investigating new concepts for a telecommunications system. The remote collaboration consisted of two people, a Ph.D. in electrical engineering and a Ph.D. in psychology; the standard collaboration consisted of three people, two Ph.D.s in psychology and one in computer science. All members of both teams had successfully collaborated in the past, completing projects of a similar nature and difficulty, but they had never collaborated together. Going into the experiment, each collaboration was judged as having a reasonable chance of success.

In addition to the preexisting differences in group size and personnel, the imposed constraint on which the collaborations differed was the communication mechanisms available to carry on their work. Two members of the standard collaboration had adjacent offices, and the third was on an adjacent floor of the same building. They had no constraints on their communication and communicated via scheduled project meetings, impromptu discussions in the hall, electronic mail, and telephone. In contrast, for the purposes of the experiment, the remote group agreed to have no face-to-face meetings in which they discussed their project. They refrained from both scheduled face-to-face meetings and impromptu discussions. Although their offices were on the same corridor and they continued to see each other casually and to have informal discussions, they refrained from discussing their joint project during these get-togethers. This restraint had the practical consequence of shifting their communication to more formal channels, such as scheduled video teleconferences or noninteractive electronic mail.

The video teleconference available to the remote collaboration consisted of a full motion, two-way, video/audio connection between their offices, supported by a shared electronic blackboard. These technologies required that a member of the collaboration had to schedule a meeting and then set up a connection with the other member to interact with him. Thus, like a genuinely remote collaboration, the opportunity for unplanned interaction was lowered although, with some effort, collaborators had at their disposal the means for high-quality interaction.

In order to understand how these collaborations proceeded, it is helpful to examine the data on two levels, process and outcome. The first level involves communication episodes, where we can compare the nature and quality of particular interactions; the second is a global level, where we can compare the outcomes of the collaborations as measured by their joint output, such as papers, programs, or artifacts. The data collected for this study consisted of records noting the date, length, and content of meetings between collaborators; video tapes and transcripts of most meetings for both the remote and standard groups; and the papers, programs, and artifacts produced by the collaborators.

Process

The first dimension for comparison is the quality of the meetings each collaboration held. One might expect substantial differences here because the remote collaboration could only hold meetings using its video/audio/data linkup, while the standard collaboration could hold meetings face-to-face. Using the transcripts from the meetings, the quality of the meetings was compared to criteria extracted from handbooks for running meetings (Doyle & Straus, 1985; Gordon, 1985). These criteria basically judge a meeting on whether the participants know why they are there and what they want from the meeting, whether during the course of the meeting some work is done toward meeting these objectives, and whether the participants are able to arrive at some common understanding of the meeting's outcome. To assess the quality of meetings by these criteria, one coder judged samples from the beginning, middle, and end of the transcriptions of each meeting and noted whether (a) participants explicitly mentioned their purposes for the meeting, (b) the content of their conversation was relevant to these purposes, (c) they summarized the meeting's accomplishments, and (d) they planned what they wanted to accomplish by their next meeting. By these criteria, and contrary to our expectations, all of the meetings—of both the remote and the standard groups—were of uniformly high quality. In particular, the remote, electronic meetings seemed as successful by these criteria as the face-to-face meetings of the standard group. This observation is consistent with the prior literature showing only minor differences between face-to-face and video-mediated meetings (e.g., Williams, 1977).

Qualitatively, the remote meetings seemed concentrated and intensely focused on work. Most of the collaborators' time was spent talking. In about 10% of the time during only two meetings, they supplemented their discussions with drawing on the video terminals to illustrate user interface concepts. This was similar to the behavior among the standard collaborators who used a whiteboard about 10% of the time during two meetings to illustrate user behavior and to outline their paper. Primarily, participants in the remote collaboration attended to and talked to each other on the video screens. Indeed, even though the video images in this teleconferencing system were not very large or of very high quality, and they did not support eye contact, they seemed to compel a polite attentiveness that was similar to the behavior found in the face-to-face meetings of the standard collaboration.

The remote and face-to-face meetings did differ in two major respects and one minor one. First, scheduled meetings in the standard collaboration were almost twice as long as the scheduled meetings in the remote group and appeared less efficient (mean length = 1 hour, 18 minutes for the standard collaboration versus 38 minutes for the remote collaboration; $t[25] = 4.12$; $p < .001$).

Second, meetings in the standard collaboration were more social and less task-focused than were the remote collaboration meetings; that is, they were filled with episodes of talk not directly related to the work at hand. Thus they were less efficient from a production point of view, but probably more effective at creating group solidarity. Anecdotes, workplace gossip, humor, and polite chit-chat all occurred relatively frequently. We examined 342 thirty-second segments of conversation from the standard and remote collaboration and coded whether they contained any social, non-task-oriented conversation. About 20% of the sampling units (38/188) in the standard collaboration were social, while only 5% of the sampling units (8/154) in the remote collaboration were social; $t(340) = 3.0$, $p < .05$.

Third, participants in the remote collaboration spent substantial time talking about the technology for their communication and work, while this was rarely an issue in the standard collaboration. The remote collaboration discussed tools for collaboration in about 14% of the sampling units, while the standard collaboration mentioned tools in only 2% of the sampling units. This difference in attention to technology may reflect the greater difficulty that members of the remote collaboration had in using their communication tools, or it may simply be an artifact of the novelty introduced by the experiment.

Outcomes

Even though the remote collaboration had "good" task-oriented meetings when examined one at a time, they had great difficulty in

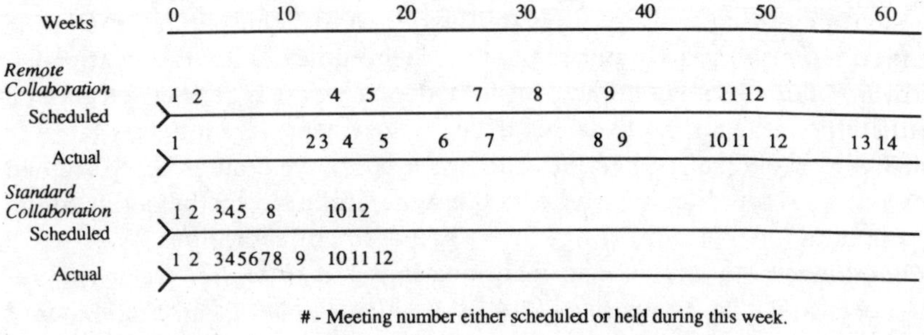

Figure 5.5. Timing of Collaborative Meetings
NOTE: For a few meetings, the initially scheduled date was not recorded in either the meeting transcript or the electronic mail.

communication when considered more globally. In particular, the remote collaborators communicated too infrequently to get their work done. Figure 5.5 shows the timing of meetings for the two collaborations over the course of their first dozen meetings. Each actual meeting and scheduled meeting is numbered chronologically. The remote group took more than twice as long as the standard group (a year versus 20 weeks) to accomplish their first dozen meetings.

Figure 5.5 also compares when meetings actually occurred with their initially scheduled dates. As can be seen, not only did the remote collaboration meet infrequently, but they were also less likely to meet close to the date they had originally scheduled to do so. For example, they had planned their second meeting to occur within two weeks of their first meeting, but it didn't actually take place until almost three months later. Thus, while the quality of each of the remote collaboration's meetings was high, their lower frequency meant that less work was accomplished in a given period of time.

If we examine meeting transcripts for the degree to which collaborators used the meetings to motivate their partners and coordinate their work, again we find few differences between the remote and standard collaborations. On this episodic level, the participants of both collaborations used their meetings to find out what their collaborators had accomplished and to motivate their partners. However, if we look at the global outcome of this effort, we find that the standard collaboration settled on the direction of its work quickly and then was able to produce a steady stream of outputs, while the remote collaboration continually redefined and reallocated its work roles. While the remote collaborators always came out of a meeting highly motivated and sure that they would fulfill their promised work assignments, they were never able to deliver any significant work outputs in the long run, and the collaboration eventually petered out.

Other Communication Modes

The collaborators had other mechanisms for communication besides scheduled meetings. The standard collaboration had impromptu meetings that it engaged in frequently; it held pairwise conversations several times per week. These impromptu meetings were either spontaneous, where a chance encounter evolved into project talk, or intended, where one collaborator sought out another to discuss a specific problem. Sometimes one collaborator would run into another in the hall, and after exchanging a few comments, they would realize that a meeting involving all team members was in order; the third partner would then be fetched to see if a near-term meeting was possible.

In addition, both groups used electronic mail. By the criteria of Figure 5.1, electronic mail falls between scheduled meetings and impromptu hallway meetings on the formality dimension. It is more spontaneous than a scheduled meeting, but less interactive. Because the remote group was prevented from using informal face-to-face conversation to motivate its work, it might very well have taken advantage of electronic mail to carry some of the coordination burden. However, excluding the exchange of document drafts, participants in the standard collaboration exchanged 190 electronic mail messages among themselves that contained about 22,000 words, while participants in the remote collaboration exchanged only 51 messages, containing about 8,000 words. On average, these numbers represent about 6.8 messages per person per month (about 1,250 words) for the standard group and 1.8 messages per person per month (about 170 words) for the remote group. On the basis of these data, collaborators in the standard group had both a higher rate of communication and longer messages than members of the remote group. Here we find some evidence that both the content of the interaction, as measured by the length of the messages, and the coordination process, as measured by the frequency of the messages, was compromised in the remote collaboration.

Interpretation

These results suggest that the suppression of informal communication opportunities between members of the remote collaboration had a negative—perhaps even fatal—effect on the work of these collaborators. Because the opportunities for informal interaction did not exist, the participants had to rely on more formal, scheduled meetings to get the work done. However, because scheduling meetings takes time and effort, a lower overall frequency of meetings took place, which in turn resulted in less work being done by the collaborators, particularly in solving the problem of refining their work plan. Thus the collaborators were not as timely in giving each other feedback on the directions their joint work should go, and therefore they took longer to construct and select among the alternatives that guided the course of their work. This problem never became clear to the participants in the remote collaboration because each meeting, in and of itself, seemed productive and useful, and the participants did not have other, more informal contacts that they could use to supplement these meetings.

In contrast, the participants in the standard collaboration used their high frequency of meetings, both formal and informal, to construct and select among alternatives quickly. The frequency of contact increased the utility of each meeting because questions that were posed could be answered quickly before too much intervening time dulled the memory regarding the point of interest. The informal contact served to reinforce decisions once they were jointly made and created meetings spontaneously if an issue had to be addressed. The velocity of the cycle of interchange among the collaborators was increased in this way, and this process served to hasten the pace of the entire collaboration.

This case study illustrates, to some degree, how formal and informal communication are used within the context of a working relationship to complement and supplement each other in driving a piece of work toward completion. Each serves a role in maintaining the social underpinnings of the relationship, and each provides some elements necessary for coordination of the production aspects of joint work.

Informal Communication and the Success of Collaborations

On the basis of a single case study, the previous section argued that more informal communication can lead to more successful collaboration. This section looks at a similar question using archival data to assess the patterns of authorship among working scientists and engineers. We attempted to determine whether physical proximity, which we view as a proxy for informal communications, was associated with successful collaboration. If a research team works together and publishes a joint paper, at some level it has solved the problems of group work, including identifying suitable partners, planning and executing work, and maintaining personal relationships.

We generated a data set based on the population of working scientists and engineers in an industrial research laboratory. The research laboratory consisted of approximately 500 Ph.D. and M.S.-level researchers in the physical, engineering, computer, and behavioral sciences. The organizational structure had three levels (laboratories, with approximately 125 members each; departments, with approximately 30 members each; and groups, with approximately 7 members each). The company was split between two campuses, located approximately 40 miles apart. Each building consisted of several floors with several wings per floor. We focused our analysis on a sample of 164 collaborating researchers, all those who had published at least two internal research reports in the preceding 30 months—at least one of these reports had to have a coauthor, and the other was either a solo-authored report or had a coauthor not included in the first report. For each of the 13,366 unique pairs of researchers (164 × 163 ÷ 2) in the sample, we obtained data on four measures: collaboration, physical proximity, organizational proximity, and research similarity.

Collaboration

Data on whether each possible pair published at least one internal research report together were obtained from a company-maintained data base of internal publications (scored 0 for no publication and 1 for a joint-authored report). Of the 164 scientists and engineers in the sample, 126 had published at least one joint article with another member in the sample.

Physical Proximity

Using the organizational phone book—which listed office addresses with codes for building, floor, and corridor—we computed a measure of physical proximity. Offices were coded 4 if they were on the same corridor of the same building, 3 if they were on same floor of the same building, but a different corridor, 2 if they were on different floors of the same building, and 1 if they were in different buildings.

Organizational Proximity

Proximity on the organizational chart was coded 4 if the pair were in th same group, 3 if they were in the same department, 2 if they were in the same laboratory, and 1 if they were in different laboratories.

Research Similarity

For each pair, we estimated the similarity of their publications on which they did not share authorship. The research similarity index is based on information retrieval techniques developed to identify semantic similarity in large text sources (Deerwester, Dumais, Furnas, & Landauer, in press). Basically, one derives the concepts in large samples of texts by computing a singular value decomposition of a text-by-word matrix, akin to factor analysis. The texts we used were abstracts of each author's noncollaborative articles. The centroid of the words in each author's abstracts was used to represent his or her work in a 30-dimensional space. The similarity of a pair of abstracts is the closeness of the concepts they contain in this semantic space. Although this is a continuous measure, for some measures we treated it categorically, differentiating the top quartile from the rest (see Kraut, Galegher, & Egido, in press, for a more complete description of the methodology).

These four measures were then examined in a number of ways. We found, for example, that researchers were most likely to achieve successful collaborations with those who were physically close to them—83% of the collaborations occurred among researchers with offices on the same floor, even though these represented only 26% of the potential collaborators. This finding is consistent with our earlier hypothesis that proximity leads to informal communication, and informal communication aids successful collaboration. However, the interpretation of this relationship is not this straightforward. The influence of physical proximity occurs partly because this organization, like many others, physically co-locates those who have needs for sustained and frequent communication; researchers in the same organization and those with research interests in common are far more likely to have offices on the same corridor than are arbitrary pairs of researchers.

Table 5.1
Association of Proximity and Collaboration

	Research Similarity				*Organizational Proximity*			
	Lowest 3/4		*Highest 1/4*		*Different Departments*		*Same Departments*	
Location	*Potential Collaborators*	*% Actually Collaborating*	*Potential Collaborators*	*% Actually Collaborating*	*Potential Collaborators*	*% Actually Collaborating*	*Potential Collaborators*	*% Actually Collaborating*
Same corridor	260	.77	333	14.41	399	6.02	194	13.40
Same floor	1755	.45	1122	4.19	2416	.70	469	8.10
Different floor, same building	3463	.06	1012	1.17	4447	.29	42	2.38
Different building	4534	.01	858	.69	5387	.13	12	0.00
Total	10023	.13	3343	3.38	12649	.48	717	9.07

A logit analysis of the bibliographic data showed that research similarity, organizational proximity, and physical proximity all independently increased the odds of collaboration. Holding constant both research similarity and organizational proximity, the coefficient for physical proximity = .613 (standard error = .143, $t = 4.26$; $p < .001$). Table 5.1 shows the relationship between proximity and collaboration in a contingency table. It demonstrates that increases in physical proximity were associated with increases in successful collaboration both for researchers in the same department and those in different departments, and both for researchers with similar and dissimilar research interests.

All the phenomena we identified above probably contribute to this association of physical proximity and successful collaboration. Researchers who have offices close to one another have many more opportunities for conversations. As a result, they are likely to know and like one another and to know of and respect one another's work. The frequent, low-cost contact made possible by physical proximity creates many opportunities for potential collaborators to become acquainted, to identify common interests, to assess interpersonal compatibility, and to do preliminary planning before they become committed to working together. In a typical collaborative relationship, neither partner starts out specifically seeking a collaborator to help carry out an already well-defined project. Rather, research projects frequently emerge from the preexisting interests and expertise of the participants in the course of casual discussion. These initial discussions became more intense and focused after researchers became committed to working together. These observations suggest that informal communication is important because it allows researchers to know about their neighbors and to develop common interests.

Once a project is started, frequent, informal communication serves as a coordination and project-management tool that helps keep the project moving forward. The coordination required of researchers becomes substantially more difficult to perform when they are not located in the same place or, as we discovered in the preceding section, when they have other constraints on their informal communication.

Summary

The preceding sections have shown that informal communication is frequent in R&D organizations, that it aids organizational members in learning about one another and their work, that it supports both production work and the social relations that underlie the work, and that it provides a critical mechanism that collaborators rely on to start joint work, maintain it, and drive it to conclusion.

Proximity leads to increased frequency of communication in general, and of informal communication specifically. Proximate colleagues have more opportunity for intended, opportunistic, and spontaneous conversations. Increased informal communication between colleagues leads to greater familiarity as well as increased respect for colleagues and their work. One would expect that the familiarity and mutual respect fostered by informal communication would be a prerequisite or at least a powerful facilitator for successful working relationships and collaborations. Colleagues need to be familiar with one another in order to seek information from and dispense information to the appropriate parties. They must be familiar with one another and share a similar perspective, context, and working culture for successful collaboration. Colleagues who collaborate must also like one another and the work that each does in order to maintain the working group and continue successful collaboration over time.

Technology for Informal Communication

In many circumstances, the ideal of close physical proximity for a working group cannot be realized. The realities of organizations may preclude individuals who are supposed to be working together from having their offices near one another. Communities of scientists, in particular, are often distributed across the nation or world. University departments often hire for breadth and may have only a single researcher covering a subspecialty. Graduate students in the job market, junior faculty denied tenure, and other faculty lured to a new institution by prestige or money frequently move far away from their advisors, colleagues, or collaborators. These factors mean that in the world of science, colleagues who would make the most suitable collaborators are often hundreds of miles away. In addition, in industry, many projects, such as software development, are so large that physical proximity for the entire staff becomes a topological impossibility. For this reason, we became interested in creating artificial proximity through the use of technology.

In the previous sections, we have seen some of the characteristics of physical proximity that make it especially suitable for informal communication. Here we review some of these characteristics and draw implications for features that artificial proximity must provide.

A Concentration of Suitable Partners

Any system that wishes to support informal interaction must provide access to a suitable population of others. In the physical world, concentration is accomplished by putting people who need to communicate close together. For example, as Table 5.1 showed, researchers in the same subunit and those who share research interests tend to be located in the same area in a building.

In the telecommunications domain, everyone who is connected to a common network is in some sense equally accessible, absent the constraints of distance-sensitive charges and unfamiliar addresses. Thus, as long as a sufficient proportion of the relevant population is connected, a population of suitable partners is also available (see Markus, 1987). Yet, because everyone is equally accessible – friends and strangers, similar and dissimilar people – the availability of suitable partners may be too diffuse to create an effective community for informal interaction. Dialing a phone number at random or posting a message to a nationwide electronic bulletin board is unlikely to put the initiator in touch with an appropriate research partner. The requirement for a telecommunications system to support informal interaction is to concentrate suitable partners.

Co-presence

For informal interaction to occur, people need an environmental mechanism that brings them together in the same place at the same time. In the physical world, this mechanism can be a lunchroom, coffee lounge, or other space where people convene, or it can involve the complexity of movement around a workplace. In the R&D laboratories we studied, people were constantly moving around. At any time, almost 50% of the staff were out of their offices, and the episodes presented previously show that unplanned interactions occur frequently during these travels. Yet much information technology, such as electronic libraries and data bases, reduces the need for physical movement to get work done, thereby reducing the opportunities for casual interaction (e.g., Kraut, Dumais, & Koch, 1989). Similarly, separating people (into branch locations, for example) reduces the likelihood that they will occupy a common physical space and thereby reduces the likelihood of informal interactions as well.

The essence of telecommunications – literally communication from afar – is co-presence without physical proximity. While most telecommunications assume intentional action to initiate communication with a particular other, it is possible to devise other idioms for nonintentional telecommunications. Electronic bulletin boards, in which readers come across the postings of other bulletin-board users, are one such example.

Low Personal Costs

The cost of communication, in terms of the amount of effort needed to initiate and conduct a conversation, is very low in the casual encounters we described previously. Often the contact is a side effect of other activities and, as such, involves no extra cost. To provide this low-cost communication when people are dispersed, we need to make getting in touch with others as easy as bumping into them in the hallway. The behavioral costs of accessing a communications system and getting in touch with the desired other party are an important determinant of that system's usefulness. If the costs of using a communication source are too high (e.g., Cullan, 1983), as is the case with many of the "traditional" teleconferencing systems (Egido, in press), the user will be either unable or unwilling to use that system for the brief, frequent, spontaneous conversations that are characteristic of informal communication.

Visual Channel

The visual channel plays an important role in informal communication. As discussed previously, seeing someone provides a means for recognizing the presence of another person, determining who he or she is, and assessing that person's availability for interaction. It also serves as a stimulus for picking a topic of conversation – reminding one person about something he or she wanted to discuss with another. Finally, the visual channel combined with the audio channel provides a medium by which the conversation is actually accomplished. A technology for informal interaction must support both audio and video communication if we are to successfully stimulate chance encounters.

These characteristics of physical proximity suggest some boundary conditions on potential telecommunication technologies to support informal communication. We use them to form the initial requirements for technologies to support informal interactions. Two prototype telecommunication systems that take different approaches to solving these problems are described below – the VideoWindow System and Cruiser.

The VideoWindow Teleconferencing System

Imagine sitting in your workplace lounge drinking coffee with some colleagues. Now imagine that you and your colleagues are still in the

Figure 5.6. Illustration of the VideoWindow System in Use

same room, but are separated by a large sheet of glass that does not interfere with your ability to carry on a clear, two-way conversation. Finally, imagine that you have split the room into two parts and moved one part 50 miles down the road, without impairing the quality of your interaction with your friends. That scenario illustrates the goal of the VideoWindow project: to extend a shared space over considerable distance without impairing the quality of the interaction among users or requiring any special actions to establish a conversation.

The VideoWindow System (Bellcore, 1989) connects two public lounge areas with high-bandwidth video channels and full-duplex four-channel audio. Figure 5.6 illustrates the VideoWindow System in use. The video images are projected onto a 3-foot-high by 8-foot-wide "picture window," roughly twice the width of usual projection TV. This allows nearly the entire lounge area to be visible, with people appearing to be about the same size as if sitting across the room. The four-channel audio provides sound localization so that the speaker's voice appears to originate from the location of the speaker's image. The system is left on 24 hours a day. To use it, a person need only walk into the room, glance at the window, and say "hello" to a person at the other end. At first glance, the VideoWindow System offers a very powerful sense of shared space and presence at a very low behavioral cost to the user.

In order to look more closely at the effects of the VideoWindow System on informal communication, we conducted a 3-month trial in our laboratory. Two commons areas on different floors of a building were connected using the VideoWindow System. Over the course of the experiment, participants were questioned about their use of the system; their knowledge of other participants, both local and remote; and their feelings about the technology. Video records were kept of face-to-face interactions in the lounge areas and of interactions that took place over the VideoWindow System.

We observed a total of 628 instances in which people were present at both ends of the VideoWindow System (364 cases of exactly 2 people, 264 cases of more than 2 people). Each instance represented an opportunity for people to strike up a conversation across the Video-Window System link. These opportunities were actually converted into verbal interaction (either a greeting or a conversation) on 153 occasions. This conversation rate of 24% represents 11% of the 2-person opportunities and 32% of the 3-or-more-person opportunities actually being used for interaction.

The videotapes of these interactions across the VideoWindow System suggest that in some cases the system was a relatively transparent medium. Many interactions that took place via the window seemed indistinguishable from similar face-to-face interactions. People often spoke and acted in a fashion that, at least superficially, seemed altered in only minor ways by the technology. They spoke a little louder and often embedded a discussion of the VideoWindow System itself in their discussion. Most of the 153 interactions we recorded would not have occurred had the system not been in place.

Yet the conversion ratio across the VideoWindow was substantially lower than the conversion ratio for face-to-face opportunities. For this comparison, we examined 81 cases in which two people were together on a single side of the VideoWindow—that is, in the same room—and thus had an opportunity to converse face-to-face. Disregarding the 29 cases in which the face-to-face people were already talking prior to entering the VideoWindow room, we found that 31% of the face-to-face opportunities were converted into interaction. This is almost three times the conversion rate for pairs across the VideoWindow.

This version of the VideoWindow System had some properties that worked against the initiation and maintenance of informal interactions. Some of the missed opportunities to converse across the VideoWindow System were due to design flaws in the user interface. About 18% of missed opportunities could be attributed directly to problems with camera and microphone placement. For example, people often tried to strike up a conversation by moving closer to the window. This caused their head to move out of the picture so that they could not be recognized, and put them out of microphone range so they could not be heard. This type of flaw could be fixed with feedback that let a communicator know whether he or she could be seen at the other side of the system. Other problems may be more endemic to the underlying technology. Eye contact is an example. Humans can shift their gaze to meet another's eyes wherever that person may be, but conventional cameras have a fixed point of view. This leads to anomalies in apparent eye contact that do not occur when people are face-to-face. Another problem is that it is easy for people to stand out of camera range where they can see the image in the window but not be seen by people at the other end. (Note the man at the left edge of Figure 5.6, who can see others in the system, but cannot be seen by them.) This contrasts with face-to-face interaction, where covert observation rarely occurs and frequently requires active concealment on the observer's part. Solving these sorts of problems requires rethinking how video cameras and monitors are designed and placed in personal communications systems.

Once a conversation was initiated, communication over the VideoWindow System sometimes broke down. One problem was that the system made it difficult to establish private conversations in otherwise public areas. In fact, the ability to make a conversation private was one of the most frequently requested improvements mentioned by users in our sample. When people in a conventional meeting room wish to exchange private information, they simply move closer together and lower their voices. This capability is not supported by the VideoWindow System or any teleconferencing systems designed to link public areas together with hands-free audio. Teleconferencing technology is designed to allow several people to converse clearly at the same time. The case of private conversations illustrates a more general point: Technology-mediated systems do not currently have the communication flexibility and ability to manipulate media characteristics that we take for granted in face-to-face conversations.

Another class of problems stems from the social and environmental context in which the VideoWindow System is situated. A VideoWindow System installation requires that users go to the room in which it is installed in order to use it. This imposes a moderate behavioral cost on users to access the technology unless they are already using the public space. The centralized nature of the VideoWindow System also results in a sampling problem: Potential encounters are limited to the set of people who occupy the system's extended space at the same time. On average, people are more likely to be someplace other than in the VideoWindow System room. In addition, the probability of being acquainted with persons at the other end may be low, given the low degree of familiarity with people at remote locations. As we have seen, even within a single building, a person is less than half as likely to know someone on an adjacent floor as someone on his or her own floor.

Although the VideoWindow System can support informal interaction at a distance and may be useful in some contexts, its implementation requires careful attention to social detail. One could imagine connecting the cafeterias of two research organizations whose members already know one another and who have reason to work together. In other circumstances (where the VideoWindow System connects strangers, and where people occupy the system locations infrequently), the probability of encountering a suitable partner and striking up a conversation is low. So, even though VideoWindow System may extend the boundaries of one's physical space and increase the absolute number of people with whom one can interact, it does not necessarily extend the effective work space or social environment in which most people work and in which most informal conversations take place.

However, even when all these problems are accounted for, the VideoWindow System was not as effective as we had expected due to factors we still do not understand. One problem may be that the mechanisms that the human perceptual system uses to judge the distance and size of objects do not operate correctly in this technology. In addition, it may be that even with life-sized images, the psychological distance to someone at the other end of a VideoWindow System link is greater than that in a comparable face-to-face situation. People had to work hard in the face-to-face situation to ignore someone who was physically in the same room. When they wanted to avoid conversation, people went to great lengths to avoid eye contact. In contrast, ignoring another person on the opposite side of the VideoWindow System was much easier, and looking at them didn't seem to create the obligation to engage in conversation. In spite of its value, then, the VideoWindow System currently does not provide the same degree of social intimacy as does face-to-face interaction.

The implications of this experiment are clear. First, we must pay close attention to the human factors of system design. Simply connecting two locations is not enough—we must also ensure that the technology can be used easily and without errors. Problems must be engineered out of the system. This is especially important with audio/video systems, because while these technologies appear to be like face-to-face communication, the subtleties of camera framing and audio placement lead to important differences that users can't be expected to understand. To solve these problems, we must learn more about how the characteristics of basic video technology affect the usability of telecommunication systems and about the difficulties caused by fixed camera positions and audio that is insensitive to user behavior. Second, we must reduce the behavioral cost to the user of gaining access to the technology in the first place. One solution to this problem is to bring the system to the user. Finally, we must address the person-sampling problem; the user must be able to get in touch with the right people regardless of location.

Cruiser

The Cruiser system (Fish, 1989; Root, 1988) is a switched telecommunication system (like the telephone) that allows a user to have an audio and video connection to any other user in the Cruiser network. It was designed to support informal communication at a distance among colleagues, while solving several of the problems identified during the VideoWindow System experiment. In particular, it lowers the behavioral cost of communication by placing audio and video equipment on the user's desktop. In addition, Cruiser provides innovative mechanisms for initiating connections between users to encourage frequent, informal, and unplanned communication among members of a distributed community. These mechanisms provide access to a large but select group of conversational partners independent of location.

Cruiser and Informal Communication

Users of this network can treat it merely as a video telephone and call any other user directly. However, Cruiser was explicitly designed to support serendipitous interactions, and a measure of its success is the extent to which it leads to fruitful or enjoyable conversations that were unintended. One mechanism for these chance encounters is the "cruise." Here, Cruiser tries to simulate the experience of walking through a hallway and stopping to chat with whomever one encounters along the way. The virtual hallway is the set of offices and other locations connected to the system. The set of locations the user actually visits is called a path; this can be one or several stops along the cruise. A single command causes Cruiser to switch to each location on the user's path in turn, pausing to give the user a brief audio and video "peek" into each office along the way. Both parties can see each other and can determine whether they wish to have a conversation and if their partner is available. If the user spots someone to chat with, another command interrupts the cruise and connects the two locations until the user decides to move on again. If the user does not wish to converse, the connection times out after a few seconds.

The path can be explicitly set by users who type a list of other users to visit. More interestingly, the path can be determined by the Cruiser system itself, by selecting from a pool of potential conversational partners. Users, a system administrator, or intelligent software can create a separate pool for each user and assign to each person in the pool a weight indicating the relative probability of that person being visited on a cruise. For example, users may list people whom they would like to see more frequently. The system administrator may add other people to this list based on such criteria as similarity of background or current work projects. In an R&D organization, for example, one might have researchers in small specialty areas "running into" one another and keeping in touch over the Cruiser system. Finally, the pool of potential partners might be updated by computer programs that monitor the environment, add to the pool, and change probabilities depending on current conditions. For example, if a member of a research organization publishes a new research report in a user's area of expertise, software might increase the weight that partner has in the user's pool. Similarly, a computer program could change the weight of individuals in a user's pool depending on the history of that user's other communication activities (i.e., telephone calls or electronic mail). For example, Cruiser might increase the opportunity for chance encounters with people one is already going out of one's way to reach.

Cruises can be explicitly initiated by typing the "cruise" command. The motivation may be work-related (e.g., "I need an answer to a question. Who is around to help me?") or social (e.g., "I'm bored. Who is free for a chat?"). In addition, cruises and opportunities to converse with other users may happen as side effects of other noncommunication-oriented activities. Just as a person might stop to converse with someone while walking down the hall to pick up a computer output, the Cruiser software itself might initiate a cruise when a user types the "print" command on a computer terminal. Linking a cruise to the reading of electronic mail or bulletin boards is especially appropriate because by executing these computer commands a user is implicitly saying, "I am interrupting my ongoing activities to be put into contact with social information."

Cruiser connections may also be initiated when two users are participating in the same event. Cruiser can connect offices with public conference rooms so that a user might attend a lecture through Cruiser. Just as conversations often break out among the audience at a lecture, such is also the case with Cruiser—one can scan and be connected to other people using Cruiser to attend the lecture. Combining special video hardware (called picture-in-picture devices) with audio switching allows such conversations to take place without disrupting the lecture.

Issues in the Trade-Off of Access and Control

We hypothesize that these technologies can be used to overcome some of the barriers to informal interaction that physical dispersion introduces. Although we have conducted a behavioral experiment with the VideoWindow System and discovered some of its limitations, at the time this book went to press, we still have little experience with the use of Cruiser.

The capabilities that VideoWindow and Cruiser provide come at a price. In particular, the value of informal communication is often asymmetrical. While we want easy access to other people, we want to control access to ourselves. Examining conventional conversations in the workplace, we see that a convenient, intentional communication for one person is an interruption for the person who is being addressed. We are discovering in building our technology prototypes that important design issues consist of resolving the trade-off between achieving access to others and protecting their time and privacy.

One of the things that makes some work settings a social environment is the relatively free access coworkers have to one another. People often gather in hallways, cafeterias, and other common areas where they are free to converse. Offices are often open to coworkers, providing accessible communication channels for both signaling and interaction. It is considered appropriate behavior to glance into an open office while walking down the hallway, and it is usually acceptable to drop in and have a chat on the spur of the moment. While these characteristics of offices are undoubtedly culturally variable (Hall, 1966) and depend on the relative status of coworkers and current trends in both architecture and job design, they are common in the relatively open and relaxed research and development environments we observed. For these mechanisms to work, people need to make themselves available. The bulk of this chapter has tried to demonstrate that people do indeed make themselves available and that, as a result, informal interaction is one of the primary communication tools of the workplace.

For telecommunication systems to support informal interaction, people need to make themselves available in this medium as well. With the VideoWindow System, public spaces must be continually connected to support drop-in conversations. Cruiser requires that users leave a camera and microphone constantly on in their offices. As a result, both of these technologies raise questions of privacy and control that pose fundamental problems for the design of telecommunication systems for informal interaction.

At the heart of the matter is the trade-off between easy access for informal communication and the ability to control one's communication environment. Users have two major concerns. First, they are worried that their privacy will be violated; others will see them and learn information about them without the user's ability to control what is imparted or without their knowledge. Users are concerned about giving off both visual and verbal information (Sundstrom, 1987). Surprisingly, most potential users of Cruiser and other video communication prototypes with similar goals (Abel, in press) believe that audio information is more revealing than video. Compounding this general concern about privacy is the worry that those with power might abuse the system by snooping surreptitiously. The second major concern of users is about uncontrolled intrusions; telecommunication provides an additional channel through which visitors can intrude into the user's work space. For all the value of informal communication, it has the drawback of consuming a scarce resource: time. Telecommunication systems like Cruiser enlarge the pool of people who can steal time from priority tasks. People of high status in organizations have developed both social and technological mechanisms, such as secretaries, appointment calendars, and answering machines, to prevent interruption. In a sense, Cruiser subverts these mechanisms. Potential users of Cruiser do not want to be open to a large number of people who intrude upon their space and take up their time.

In the physical environment of the office, these matters are partially regulated through social conventions that dictate who can interrupt whom, when to knock, and so on. In a public space, a conversation may be made private by the participants lowering their voices or moving to a private space. Office occupants may exert direct and explicit control over their local environment, for example, by closing the door to the office and refusing to answer their phone. These simple acts create barriers to the flow of information in or out of the office and signal to potential visitors about the occupant's availability for interruption.

These considerations have led us to design a number of features for providing users with control over their privacy and access. First is the use of reciprocal views, or the "see and be seen" strategy. When a person visits another office using Cruiser, two things happen: The visitor can see and hear what is happening in the other office, and the occupant of the visited office also sees and hears what is happening in the visitor's office. This feature ensures that occupants are not spied upon or observed without their knowledge. Next, in order to give users the ability to look into an office to ascertain the occupant's presence or simply glance in while passing by, we have adopted the notion of a limited duration "peek" (Goodman & Abel, 1987), still following the principle of reciprocal viewing. This feature provides sufficient time to see if the occupant is available but limits the duration of an intrusion. Third, we have heard from many people that one advantage of the real office is that they can hear someone walking past the doorway. We simulate this in Cruiser by providing an audible signal when someone peeks into a user's office.

Finally, we can provide a number of controls to give occupants the functional equivalent of a door between them and Cruiser's virtual hallway. Occupants have the option of setting a busy flag on the audio and/or the video signals emanating from their offices. When a busy flag is set, visitors will receive a notice on their workstation that a channel is not available. If, for example, the video is open but the audio is busy, a visitor will still get the short visual peek into the office but will not be able to hear anything. If an interaction ensues, the occupant may, of course, allow audio communications at any time. These privacy flags may either be set explicitly with a workstation command or implicitly by sensors attached to the physical door or telephone so that the busy flag is set when the office door is shut or the telephone is in use.

A more sophisticated possibility is based on the existence of a hierarchy within the workplace that gives certain people greater privileges to interrupt than others. In general, those with higher status and those with important business (secretaries, for example) have greater freedom to interrupt than peers or lower status individuals. The system already knows who is visiting whom. The addition of organizational and status information can let it exercise a modicum of "intelligence" in deciding when a visitor might interrupt a conversation or override a privacy flag. Balancing the interruption privileges of high-status visitors against the privacy concerns of other individuals requires sophisticated filters and careful design in order to prevent abuses and maintain user confidence in the technology. If a rich simulation of office etiquette makes the technology more effective, however, then more advanced software techniques from the artificial intelligence domain could be utilized.

Many of these features—selective media blocking, reciprocal views, visual status cues, and user-specific filters—will be incorporated into Cruiser and evaluated in field trials, but, again, there is a design trade-off: At what point do the protection schemes increase the cost of communication to the point where drop-in interactions will simply not occur, thereby defeating the goals of the system? One intent of our experimental trials is to reveal whether these manipulations are either necessary or sufficient for users' acceptance of the system.

Conclusions

The intricate choreography necessary to do work in organizations requires effective coordination. Throughout this chapter, we have argued that informal communication, generally mediated by physical proximity, is crucial for this coordination to occur. Informal encounters are useful means of getting people to know and like one another, of creating a common context and perspective, and of supporting planning and coordination in group work. Indeed, without them, collaboration is less likely to start and less productive if it does occur. Physical proximity helps by allowing appropriate people to encounter one another frequently, by supporting visual channels to induce and assess readiness for communication, and by supporting highly interactive conversation.

What happens when groups get too large or spread out to allow physical proximity to support informal communication? Telecommunication and computer technology may be able to take the place of proximity. In this light, we presented the VideoWindow System and Cruiser. The VideoWindow System is based on the concept of extending a large shared space and providing informal communication access via video and audio. The system is available 24 hours a day, and workers merely enter the room, "bump" into someone on the other end, and converse. Cruiser is based on the metaphor of walking along a hallway and stopping to talk with whomever is encountered along the way. Cruiser provides informal communication access by putting the audio and video equipment on the user's desktop and by using a switched audio-video network to connect offices and other locations together, leading to a large number of potential connections.

We have discovered that designing telecommunication systems to support informal interaction at a distance requires the resolution of many trade-offs. System designers must understand and be sensitive to the needs and concerns of system users. They must be alert to the subtleties of etiquette and the protocols that govern social interactions, be concerned with the possibility of unwanted intrusions or surveillance, and balance the need for casual access against the desire for control of one's personal space. A system must provide access and openness as well as restrictions and privacy. There is a need for mechanisms that support subtle communication protocols and are sensitive to the social context in which communication systems are embedded. We need to explore these issues with real prototypes and users, and we need to develop appropriate design and evaluation methodologies for innovative system design and effective evaluation. It remains to be seen whether we will be able to solve these problems in the VideoWindow and Cruiser systems.

Notes

1. Sunita Ashar and Jonathon Shulman served as able research assistants in this data collection.

2. Steven Rohall wrote the program to compute the distances.

References

Abel, M. J. (in press). Experiences in an exploratory distributed organization. In J. Galegher & R. Kraut (Eds.), *Intellectual teamwork: The social and technological bases of cooperative work.* Hillsdale, NJ: Erlbaum.

Allen, T. (1977). *Managing the flow of technology.* Cambridge, MA: MIT Press.

Argote, L. (1982). Input uncertainty and organizational coordination in hospital emergency units. *Administrative Science Quarterly, 27,* 420-434.

Bellcore, or Bell Communications Research. (1989). *The VideoWindow Teleconferencing Service model* (Special Report SR-ARH-001424). Morristown, NJ: Author.

Berscheid, E., & Walster, E. H. (1969). *Interpersonal attraction.* Reading, MA: Addison-Wesley.

Blau, P. M., & Scott, W. R. (1962). *Formal organizations.* San Francisco: Scott, Foresman.

Bossard, J. (1932). Residential propinquity as a factor in marriage selection. *American Journal of Sociology, 38,* 219-224.

Brown, P., & Fraser, C. (1979). Speech as a marker of situation. In K. Schere & H. Giles (Eds.), *Social markers in speech* (pp. 33-62). New York: Cambridge University Press.

Clark, H., & Marshall, C. R. (1981). Definite reference and mutual knowledge. In A. K. Joshi, B. L. Webber, & I. A. Sag (Eds.), *Elements of discourse understanding* (pp. 10-63). Cambridge: Cambridge University Press.

Cullan, M. J. (1983). Environmental scanning: The effects of task complexity and source accessibility on information gathering behavior. *Decision Science, 14,* 194-206.

Daft, R. L., & Lengel, R. H. (1984). Information richness: A new approach to managerial behavior and organizational design. In B. Staw & L. L. Cummings (Eds.), *Research in organizational behavior* (Vol. 6, pp. 191-233). Greenwich, CT: JAI Press.

Daft, R. L., & Lengel, R. H. (1986). Organizational information requirements, media richness, and structural design. *Management Science, 32,* 554-571.

Deerwester, S., Dumais, S., Furnas, G., & Landauer, T. (in press). Indexing by latent semantic analysis. *Journal of the American Society for Information Systems.*

Doyle, M., & Straus, D. (1985). *How to make meetings work.* New York: Jove.

Egido, C. (in press). Teleconferencing as a technology to support cooperative work: Its possibilities and limitations. In J. Galegher & R. Kraut (Eds.), *Intellectual teamwork: The social and technological bases of cooperative work.* Hillsdale, NJ: Erlbaum.

Festinger, L., Schacter, S., & Back, K. (1950). *Social pressures in informal groups: A study of human factors in housing.* Stanford, CA: Stanford University Press.

Fish, R. S. (1988, February). *Comparison of remote and standard collaborations.* Paper presented at the Workshop on Computer-Supported Cooperative Work, Tucson, AZ.

Fish, R. S. (1989). Cruiser: A multimedia system for social browsing. *SIGGRAPH Video Review* (Video Cassette), Issue 45, Item 6.

Gabarro, J. J. (1987). The development of working relationships. In J. W. Lorsch (Ed.), *Handbook of organizational behavior* (pp. 172-189). Englewood Cliffs, NJ: Prentice-Hall.

Goffman, E. (1963). *Behavior in public places*. New York: Free Press.

Goodman, G. O., & Abel, M. J. (1987). Communication and collaboration: Facilitating cooperative work through communication. *Office: Technology and People*, *3*, 129-146.

Gordon, M. (1985). *How to plan and conduct a successful meeting*. New York: Sterling.

Granovetter, M. (1973). The strength of weak ties. *American Journal of Sociology*, *68*, 1360-1380.

Hackman, J. R. (1987). The design of work teams. In J. W. Lorsch (Ed.), *Handbook of organizational behavior* (pp. 315-342). Englewood Cliffs, NJ: Prentice-Hall.

Hagstrom, W. O. (1965). *The scientific community*. Carbondale: Southern Illinois University Press.

Hall, E. T. (1966). *The hidden dimension*. Garden City, NY: Doubleday.

Kendon, A., & Ferber, A. (1973). A description of some human greetings. In R. Michael & J. Crook (Eds.), *Comparative ecology and behavior of primates* (pp. 591-668). London: Academic Press.

Krauss, R. M., & Fussel, S. (in press). Mutual knowledge and communicative effectiveness. In J. Galegher & R. Kraut (Eds.), *Intellectual teamwork: The social and technological bases of cooperative work*. Hillsdale, NJ: Erlbaum.

Kraut, R. E., Dumais, S., & Koch, S. (1989). Computerization, productivity and quality of work-life. *Communications of the ACM*, *32*, 220-238.

Kraut, R. E., Egido, C., & Galegher, J. (in press). Patterns of contact and communication in scientific research collaboration. In J. Galegher & R. Kraut (Eds.), *Intellectual teamwork: The social and technological bases of cooperative work*. Hillsdale, NJ: Erlbaum.

Kraut, R. E., Galegher, J., & Egido, C. (1988). Relationships and tasks in scientific collaboration. *Human-Computer Interaction*, *3*, 31-58.

Kraut, R. E., Galegher, J., & Egido, C. (1989). *Informal communication in scientific work*. Unpublished manuscript, Bell Communications Research, Morristown, NJ.

Kraut, R. E., Lewis, S. H., & Swezey, L. W. (1982). Listener responsiveness and the coordination of conversation. *Journal of Personality and Social Psychology*, *43*, 718-731.

Maisonneuve, J. A., Palmade, G., & Fourment, C. (1952). Selective choices and propinquity. *Sociometry*, *15*, 135-140.

March, J. G., & Simon, H. A. (1958). *Organization*. New York: John Wiley.

Markus, L. (1987). Toward a "critical mass" theory of interactive media: Universal access, interdependence and diffusion. *Communication Research*, *14*, 491-511.

McGrath, J. E. (1984). *Groups: Interaction and performance*. Englewood Cliffs, NJ: Prentice-Hall.

McGrath, J. E. (in press). Time matters in groups. In J. Galegher & R. Kraut (Eds.), *Intellectual teamwork: The social and technological bases of cooperative work*. Hillsdale, NJ: Erlbaum.

Mintzberg, H. (1973). *The nature of managerial work*. New York: Harper & Row.

Monge, P. R., & Kirste, K. (1980). Measuring proximity in human organizations. *Social Psychology Quarterly*, *43*, 110-115.

Ouchi, W. G. (1980). Markets, bureaucracies, and clans. *Administrative Science Quarterly*, *25*, 129-141.

Over, R. (1982). Collaborative research and publication in psychology. *American Psychologist, 37*, 996-1001.

Pelz, D. C., & Andrews, F. M. (1966). *Scientists in organizations.* New York: John Wiley.

Reder, S., & Schwab, R. (1988). The communicative economy of the workgroup: Multichannel genres of communication. In *Proceedings of the 1988 conference on computer-supported cooperative work* (pp. 354-368). New York: ACM Press.

Root, R. W. (1988). Design of a multimedia vehicle for social browsing. In *Proceedings of the 1988 conference on computer-supported cooperative work* (pp. 25-38). New York: ACM Press.

Segal, N. W. (1974). Alphabet and attraction: An unobtrusive measure of the effect of propinquity in a field setting. *Journal of Personality and Social Psychology, 30*, 654-657.

Sproull, K. (1984). The nature of managerial attention. In L. Sproull & P. Larkey (Eds.), *Advances in information processing in organizations* (pp. 9-27). Greenwich, CT: JAI Press.

Stohl, C., & Redding, W. C. (1987). Messages and message exchange processes. In J. Jablin, L. Putnam, K. Roberts, & L. Porter (Eds.), *Handbook of organizational communication* (pp. 451-502). Newbury Park, CA: Sage.

Suchman, L., & Wynn, E. (1984). Procedures and problems in the office. *Office: Technology and People*, *2*, 113-154.

Sundstrom, E. (1987). Privacy in the office. In J. D. Wineman (Ed.), *Behavioral issues in office design* (pp. 177-201). New York: Van Nostrand Reinhold.

Tajfel, H. (1970). Experiments in intergroup discrimination. *Scientific American, 223*, 96-102.

Tushman, M. L., & Nadler, D. (1978). Information processing as an integrating concept in organizational design. *Academy of Management Review, 3*, 613-624.

Van de Ven, A. H., Delbecq, A. L., & Koenig, R., Jr. (1976). Determinants of coordination modes within organizations. *American Sociological Review, 41*, 322-338.

Williams, E. (1977). Experimental comparisons of face-to-face and mediated communications: A review. *Psychological Bulletin, 84*, 963-976.

Zajonc, R. B. (1968). Attitudinal effects of mere exposure. *Journal of Personality and Social Psychology, 9*, 1-27.

Zipf, G. K. (1949). *Human behavior and the principle of least effort.* Cambridge, MA: Addison-Wesley.

PART III

AEROSPACE

6

Crew Factors in the Aerospace Workplace

BARBARA G. KANKI
H. CLAYTON FOUSHEE

Ever since Tom Wolfe's (1979) insightful and irreverent social documentation of the first U.S. astronauts, "the right stuff" has come to represent a familiar stereotype of the white-scarf-and-goggles pilot who maintains his cool demeanor as obstacles and risks mount to staggering proportions. Even today, a member of the pilot "brotherhood" (whether in aviation or space) is often depicted as a "fearless, self-sufficient, technically qualified, and slightly egotistical individual, whose job description calls for the defiance of death on a regular basis" (Foushee & Helmreich, 1988, p. 191). While exaggerated, or perhaps simply outdated in many of today's aerospace workplaces, such stereotypes can be traced to legitimate origins. For example, Wolfe writes,

> As late as 1970, I was to discover in an article by a military doctor in a medical journal, a career Navy pilot faced a 12 percent likelihood of dying in an accident. This did not even include deaths in combat, which at that time, with the war in Vietnam in progress, were catastrophically high for Navy pilots. The Right Stuff became the story of why men were willing—willing?—delighted—to take on such odds. (1979, pp. x-xi)

However, navy pilots were the epitome of the rule and not the exception. In the earlier days of aviation, single-pilot flight control was not only the primary task, but it was one fraught with equipment unreliability; the technology of that era provided much thinner insulation from unfriendly skies (i.e., darkness and poor weather). In short, the piloting profession was filled with real dangers that required quick-thinking ingenuity, and pilots often flew "by the seat of their pants" to get the job done. Not surprisingly, the "right stuff" attributes would thrive in this earlier environment. Furthermore:

> These characteristics were much admired and became not only essential elements of the informal pilot culture, which were passed from generation to generation of pilots; but they were also institutionalized in regulations governing pilot performance, as well as in policies guiding pilot selection criteria. They are so much a part of the culture, that they are still with us in an age where flight operations are routine and aviation is by far the safest mode of commercial transportation. (Foushee & Helmreich, 1988, p. 191)

The environment in which the aviation culture was born was not to endure for very long. As in many workplaces, technological changes and growth led to many successful innovations, but a mismatch grew between workers, tasks, and the workplace. Many advances were made, but the system as a whole did not advance at an equal pace. As a result, problems of human error began to affect crew performance in a variety of ways at both the individual and group levels. These problems occurred with respect to the environment and task design, as well as in organization and management structure. Aspects of training and intervention also became increasingly less well suited to the aviation workplace.

In this chapter we first examine the relationship between technological change in the aerospace workplace and the general problems listed above. Next, a philosophy and conceptual framework for conducting research dealing with these problems are described, and several specific studies are described in detail. Finally, a summary of the operational products we expect to result from research on these problems concludes the chapter.

Effects of Technological Change in the Aerospace Workplace

Advances in technology have revolutionized the aviation workplace (commercial, military, and general aviation operations) in a great variety of ways. Because our focus is on crew performance, we are primarily concerned here with multipilot operations as opposed to single-pilot ones. Each pilot's task has changed dramatically, yet these changes have not always been accompanied by a concern with how individual task changes fit into the integrated system of changes or, more simply, how changes in one task affect other tasks. Omissions of this type have major implications, not only in daily aviation operations, but in the company policies, standard procedures, and training programs that underlie them. Superficial changes are often presumed to be sufficient solutions until accidents and incidents prove otherwise—that is, until the problems become visible and the damage has been done. In many ways, the pilot's workload has decreased due to advances in technology, thereby creating the illusion that there is less risk of human error. The overall aviation safety record, plus the fact that many problems related to mechanical reliability *have* disappeared, reinforces that notion. The volume of air transport continues to increase, and operations take on an utterly routine look. Even the space industry appeared to be settling into an attitude of routineness until the *Challenger* disaster dispelled that illusion in 1986.

The role and task of the pilot have changed due to: (1) overall improved equipment reliability, (2) an exponential increase in the number of controls and displays available to the pilot, coupled with a proportional increase in aircraft functions and performance (see Baron, 1988; Sexton, 1988), (3) the development of multipiloted aircraft and systems of redundancy as a response to increased function and workload, (4) increasingly automated systems, which often involve computer-based devices (see Wiener & Curry, 1980), (5) the introduction of new warning and alerting systems, and (6) "the increasing complexity of the environment in which pilots fly—an environment congested with aircraft, demands, regulations, and procedures" (Wiener, 1988, p. 451).

Changing Roles and Increasing Human Error

While the consequences of technological changes include a net gain in overall safety, there has been a pronounced shift in problem areas over the decades—namely a general decrease in mechanical problems but an alarming increase in the proportion of human error problems. The International Civil Aviation Organization (1984) noted that as aircraft and systems have become more reliable over time, the relative proportion of accidents attributable to human error has systematically increased. In recent years, even the most conservative estimates attribute 65% of all aircraft accidents to the flight crew (Foushee, 1984). For example, a study of the worldwide commercial jet fleet, conducted for Boeing by Lautman and Gallimore (1987), concluded that while "the air safety record has improved substantially in the decades since the introduction of the jet transport in 1959 . . . crew-caused accidents dominated all other causes by a wide margin. It has consistently accounted for over 70% of the major (fatal or hull loss) accidents" (p. 1). Between 1959 and 1985 (N = 310 hull loss accidents with known causes, excluding sabotage and military action), primary cause factors included the airplane (11.6%), maintenance (1.6%), weather (5.5%), airport air traffic control (4.2%), and a miscellaneous category (3.2%), in addition to flight crew (73.9%).

Public and company awareness of this problem has created pressure to reconsider both pilot selection criteria and proficiency standards. However, little has changed since the work of human factors psychologists dating back to World War II, in which efforts were focused on the identification of technical or "stick and rudder" skills, and attributes of pilots as individuals (e.g., personality characteristics related to perceptual skills, or emotional stability required for operations in stressful circumstances). See Ellis and Conrad (1948) and Dolgin and Gibb (1988) for reviews of this research.

Although proficiency standards and the accompanying selection criteria were developed largely for single-pilot operations, they were simply extended to multipilot operations, disregarding the fact that sophisticated multipilot transport aircraft require a highly structured group performance in which a number of interpersonal factors are relevant to crew effectiveness. Thus we have been caught in a serious time lag in which pilots are often selected on the basis of criteria corresponding to the single-pilot role and then placed in a multipilot environment. While a few high performance, single-seat pilot positions still remain (such as air-to-air combat), multipilot transport assignments are by far more typical. This dilemma has been sustained both formally by the system (in terms of selection, training, and standards), as well as informally as the "right stuff" culture is perpetuated by its members.

Inadequate Countermeasures to Task Demands

Concurrent with the development of multipiloted aircraft, the philosophy of redundancy was adopted both in flight systems and with respect to crew function. This resulted in three major consequences for the pilot's role: (1) tasks became more specialized since the explicit delegation of tasks to positions became necessary due to multiple crew positions; (2) activities required more conscientious coordination, because crew members needed to be sure that information was successfully transferred from one to another, and (3) new crew procedures were created, such as cross-checking and monitoring, that became standard operating procedures (i.e., regulations). Crew members were required to monitor one another as well as flight systems information. Consequently, what was previously a more direct, active control of the flight systems by one person became a task divided across crew members, thereby becoming more of a passive and indirect function.

While redundant systems were designed to provide a safeguard against human error, they often failed because the additional step of transferring information among flight crew members was forgotten or incompletely carried out. Ironically, cross-checking redundancy is often needed during emergency or abnormal procedures, when routines are strained by high workload and frequent interruptions. Sometimes the captain fails to listen to the warnings of subordinate crew members; sometimes the crew members fail to inject critical information into the decision-making process; and sometimes all crew members are simply engrossed in their own interpretation of the situation or their own individual tasks. Usually, it is pointless to assign blame to one person or another because the information transfer problem is inherently an interactive process. In many cases, crew members are trying to anticipate each other, but their good intentions fail. Numerous reports of these types of small failures (which can nonetheless result in large problems) can be found in the Aviation Safety Reporting System (ASRS) data files.[1] For example:

> On takeoff from Atlanta, our flight used an improper heading for climb-out. . . . During this period, immediately after liftoff and while engaged in the climb-out and weather analyzing, I said "075, the heading, right?" He (the first officer) looked at me quizzically and said what I understood as, "Yeah, OK." We continued to climb and were told to contact departure . . . they told us to turn immediately to 110 . . . then asked, "What heading were you cleared to?" I said, "Tell them 075 degrees, that's what we read back wasn't it?" The copilot did not answer me, so I looked at him and he again had that odd look. I repeated louder, "Tell them 075. . . ." At no time were we aware of the serious problem with the other aircraft that unbeknownst to us had taken off on runway 8. . . . I believe it was solely due to poor cockpit communication. . . . I thought 075 degrees was the correct departure heading, and to confirm it, I asked the copilot, but my question came at a time when we were very busy. . . . He thought I was asking his evaluation of a direction. Coincidentally, that, to him, was a good direction and he answered in the affirmative. I took his answer as a concurrence of my question of proper takeoff heading (Foushee & Manos, 1981, p. 67)

This is only one example of many that demonstrates that individual pilot performance and technical proficiency are not always enough. An analytical review of 60 air transport accidents between 1968 and 1976 that implicated crew coordination (Cooper, White, & Lauber, 1979) revealed that poor human resource management resulted from: (1) preoccupation with minor mechanical problems, (2) inadequate leadership, (3) failure to delegate tasks and assign responsibilities, (4) failure to set priorities, (5) inadequate monitoring, (6) failure to utilize available data, and (7) failure to communicate intentions and plans. Similarly, in one of the first high-fidelity simulator studies, Ruffell Smith (1979) found that most problems were related to breakdowns in crew communication and leadership, as opposed to a lack of individual technical knowledge or skill regarding navigation, systems operations, flying, or tactical decisions. Often communication and crew integration errors involved problems in information transfer, resulting in a poor distribution of work load or faulty decision making. Researchers and designers seeking solutions to human factor problems in aviation will need to acknowledge that the interface between the crew members who carry out interrelated tasks may be just as critical an aspect of the task environment as the traditional human-machine interface.

Aerospace Workplaces: Universal Problems in a Unique Environment

The concepts of systems vulnerability and human fallibility are certainly not unique to the aerospace environment, and it would be unfair to suggest that there is a *simple* causal relation between changing technology and the problems described above. Rather, the emergent problems appear to stem from a mismatch in pacing—that is, technological change in the aerospace workplace has been so great and so rapid that there has neither been the time nor the necessary guidance for making appropriate changes in the social/organizational structure of the workplace. Like human error, resistance to change is a universal problem; while it is relatively easy to make isolated or superficial technical changes, the underlying social changes accompanying technological advances are often complex and potentially traumatic to the workers in the workplace.

In the following section, specific problem areas that have developed in the aerospace workplace due to changing technology are discussed, but it should be noted that many of these problems are found in factories, offices, other systems of transportation, and a wide variety of workplaces in which individuals coordinate their activities in a common endeavor. Each environment generates its own definitions of "good" versus "poor" team performance, and each environment also generates its own primary and specific research goals. Nevertheless, general research strategies and more specific methodologies should be generalizable from one field setting to the next, as should some of the more general concepts and findings.

In spite of the universality of many of the problem areas, the aerospace workplace is also distinctive. First, the consequences of crew error on the flight deck transcend ordinary costs because often many human lives are at stake. Because tolerance for error is very low, a successful group performance must be narrowly defined. For example, if a company's primary goal is increased productivity, an increase of almost any size might be labeled a "success." However, when safety is the bottom line, only zero accidents constitute a true success. Attainment of this goal may not represent a statistically significant improvement, but it may be the difference between life and death. In this environment, seemingly small problems and miscommunications can result in great tragedy. For example,

> An L-1011 went down at night in the Everglades in 1972 killing 99 passengers and crewmembers. The accident sequence was initiated by a burned-out light bulb in the system indicating that the landing gear was down and locked; a problem which occupied all three crewmembers' attention to the extent that no one noticed that the autopilot had become disengaged. The sad result was that the aircraft slowly drifted down into the swamp without anyone noticing. An ironic part of the accident scenario was that the air traffic controller noticed on his radar display that the aircraft was losing altitude and queried the flight crew by asking "how are things comin' out there?" The crew, of course, probably thought that he was inquiring about the landing gear problem and responded by saying that everything was all right. In short, the accident probably would have been prevented had the controller simply directed his observation toward the loss of altitude in a precise manner, but of course, it *could* have been prevented in many other ways. Nonetheless, the most interesting aspect of this accident is that *no one* was actually flying the aircraft, and the system of redundancy (both in the air and on the ground) . . . was nonexistent. (Foushee & Helmreich, 1988, pp. 194-195)

The actual verbal miscommunication that took place in this major disaster was very commonplace, the problem (of losing altitude) was easy to fix, and the critical information was available. Yet, the team failed (see National Transportation Safety Board, 1972).

As researchers (and paying customers) in the aviation workplace, many people have a pretty accurate idea of the dangers, risks, and consequences of human error on the flight deck. In fact, few other workplaces receive as much public attention as commercial aviation—not even other perceived high-risk environments, such as nuclear power plants and toxic waste operations. In the space domain as well, the public is sadly aware of the problems that can occur during the prelaunch and launch phase of operations. However, only a few individuals have had direct experience with the kinds of stresses that will affect performance in long-duration isolation and confinement in space, and researchers in this area are reduced to piecing together what is known from shuttle operations, Skylab missions, and publicly available reports of Soviet operations.

> Starting with the international Space Station and continuing with excursions to the moon, Mars, and beyond, future missions will have crews of at least eight members and will extend from three to 36 months, on upwards.
>
> Crews will differ in other ways besides size. They will be more heterogeneous and will include more non-astronauts coming from a variety of scientific and industrial disciplines. Because of the extended time periods and greater distances from earth, crews will be more autonomous and will have to make more decisions affecting mission accomplishment, as well as their own safety, and well-being. The differences combined—size of crew, time together in space, increased heterogeneity, and greater decision making responsibility—will present circumstances and problems unlike anything experienced before. . . . The complicating feature is that members will not be able to get away from each other for the duration of the mission. (Nicholas, Foushee, & Ulschak, 1988, p. 1)

With the exception of prisons and some hospital settings, even the most closely tied groups, such as families and work groups, have the freedom to break up when tensions run high. Just the knowledge that a situation is temporary seems to take a tremendous strain off the members. Also, a professional attitude shown by group members can go a long way toward defusing a situation and preventing decrements in task performance; relying, however, on some members to continually compensate for the difficult few is probably only realistic for limited amounts of time or under specific circumstances where they know they can walk away at the end of the day. When the workplace is also one's living space and the only opportunity to "get away from it all" may be weeks, months, or years away, it is clear that a person's usual tolerance levels and coping mechanisms simply may not work.

It should be evident that both aeronautical and space environments have some unique characteristics but also some differences from each other. In many ways (as discussed later), research strategies, methodologies, and instruments for studying these environments must be highly specialized in order to provide both experimental control and generalizability. Yet, the problem areas themselves provide a common focus for researchers, and the following section describes some of the ways in which these problems manifest themselves in the aerospace workplace.

Problem Areas

Problems related to the introduction of new technology can be viewed from various perspectives. Considering "new technology" as relatively specific hardware or software innovations, problems can refer to specific human-computer interface tasks, to the more general human versus automation distribution of tasks, or to humans working together in a hybrid environment of manual and automated systems. However, "new technology" can be understood more globally as referring to the whole human-machine interface (e.g., a pilot flying a particular aircraft) or, more appropriately, the aircrew-machine interface – that is, how the team as a whole interfaces within the aircraft task environment. In the aviation environment, many of these kinds of problems can be directly investigated, while in the space domain, the primary objective is to anticipate the kinds of problems that crews will face and to provide guidelines for the designers of the systems.

The perspective taken in this chapter is the more global one, although human-computer interaction will be discussed as a subtopic within the problem area focusing on task issues. To present a broad view of the aerospace workplace, five problem areas will be discussed that represent different types of social psychological issues and that suggest different levels of analysis from a methodological standpoint. They are (1) individual and physiological, (2) crew and interpersonal, (3) environmental and task, (4) organization and management, and (5) training and intervention problems.

Individual and Physiological Problems

This problem area is not mentioned as a social psychological research area in itself; that is, it is not a recommendation to social psychologists to switch into areas of individual differences, learning, and biopsychology. Rather, it is included because there are extremely important individual and physiological factors that must either be held constant when focusing on group factors or whose significant impact on group function can be varied as a conditional factor. For instance, when crew performance on the flight deck is studied, we must be able to assume that individual competence and skill level is constant. This is seldom a problem because both pilots and astronauts are very rigorously selected and overtrained in technical skills. We can assume that individual differences of this type are not systematically affecting research findings due to the nature of the subject population.

On the other hand, such factors as fatigue and other stresses on individuals can in turn affect the entire crew. Personality differences, too, may generate variations in team performance, depending on how closely and how long team members work together. It is likely, for instance, that the personality of persons in authority or command will influence the way in which the team is managed. In space, of course, physiological factors become increasingly important, with the effects of microgravity playing a major role. Crews may also suffer the impact of abnormal temporal variations in day/night cycles, which can generate circadian dysrhythmia as well as mundane problems in work/rest scheduling. Both the physiological and psychological aspects of the experience of living in a confined, isolated environment in long-duration spaceflight are likely to create new and acute stresses on crew behavior. Thus, while individual differences are not the direct focus of "crew factors" research, it would be a terrible mistake to overlook their potential impact on the interactions within a group.

Crew and Interpersonal Problems

As discussed earlier, the almost exclusive focus on individual, technical qualifications of aviators has neglected the interpersonal skills that enable individuals to work more productively as a team. Disturbing rates of human-error-related accidents have motivated a more thorough investigation of the group processes relating to problem solving, decision making, information transfer, and effective leadership. Obviously, a balance must be maintained between technical qualifications and interpersonal competence, and much depends on the specific work demands. The commercial aviation flight deck, where crew members can leave at the end of a day, is a situation in which demands for performance quality are restricted to a particular time and place, and the degree of coordination required varies according to the phase of flight. In a space environment, acceptable performance may require a broader definition than productivity and safety. When members must live together for an extended amount of time, the low tolerance for dissatisfaction may require a greater degree of compatibility among crew members.

The interface between crews will also be important as task specialization increases. Already there are pilots, mission specialists, and payload specialists in space shuttle operations, and when space station operations commence, there may be science crews whose members come from a great variety of international backgrounds. With increasing heterogeneity, the potential for miscommunication, competition, and conflict may increase due to professional and cultural differences, and sometimes simply because different teams have conflicting task priorities and goals. Even aeronautical crews can have conflicting goals—for instance, when the pilots favor a strategy that simplifies their task but overloads the flight engineer. Crew interface problems are tied to the roles that individuals and subteams are assigned, and these assignments are typically dictated by aspects of the task structure, which is discussed in the next section. The consequences when previously human roles are filled by automated functions become an additional complication to the existing role and task structure.

Environmental and Task Problems

This area includes the traditional human-computer interface area, as well as the more general issue of how to distribute human and automated task control. In aeronautics this problem is compounded by difficulties occurring when switching from one mode of operation to another, or when operating in split mode (part manual and part automated). Unfortunately, it is during abnormal procedures (when the work load becomes high) that the benefits of automation are also compromised by the need for greater crew coordination. An additional complication is that aircraft equipped with varying levels of technology are occupying a common airspace and must deal with each other and with air traffic controllers whose systems may or may not match. Again, the benefits of advanced technology may be partially lost because the complex flying environment in which it operates is not wholly compatible.

As mentioned earlier, the replacement of direct flight control with more highly automated systems has changed the pilot's task in numerous ways. While pilots seem genuinely to like the new aircraft, their enthusiasm is tempered with new anxieties (see Curry, 1985; Wiener, 1985). Will critical discussions be implemented without their consent? How smoothly will a pilot make the transition back to manual controls when automation fails? Will the system eventually "lock out" the user, as car manufacturers have locked out the consumer by making certain functions (like a simple tune-up) inaccessible?

An interesting human-computer interface problem found in space shuttle operations is the astronaut implementation of telerobotics. As distinct from the more typical case of human-robotic interface, an astronaut performing EVA (extra-vehicular activity) may need to coordinate his/her activity with a human operator who is monitoring the task from within the spacecraft or, in some cases, from the ground. Although the human-machine interface is likely to be thoroughly tested in a part-task mockup, other aspects of the task execution (e.g., communication between human operators) may be trivialized in this artificial setting. For example, if the situation fails to incorporate the full complexity of the situation, such as constraints (due to scheduling, shared resources, multi-tasking, and fatigue), communication problems and delays, and other external factors that can throw the procedure off its intended track, the interface between human operators may be simplified to the point where the simulation no longer provides a valid assessment of the procedure. When the task is actually performed in space, there may be numerous communication and coordination problems among EVA and IVA (intra-vehicular activity) operators, and with technical or science support from mission control. The team, as a whole, may find itself unprepared to troubleshoot unanticipated problems on-line and interactively.

There are also problems associated with the environment in general as opposed to specific task parameters. For instance, there is a simple lack of environmental control and variability in space. In other words, there are simply fewer choices available to individuals for moving from one setting to another and a smaller range of environmental properties, such as appearance, temperature, lighting, spatial dimensions, and noise level, that can be varied. Living in total dependency on life-support systems—where direct control, if it exists at all, may be extremely limited—can create new stresses for which there are no easy countermeasures. For the space environment, it is especially important to anticipate as many problems as possible simply because there may be no feasible means of intervening once a mission is underway. Thus, the design phase of mission planning is critical. However, in both aeronautical and space workplaces, the way in which crew members relate to one another and work together can be largely influenced by the rules, roles, and procedures predetermined by the organizational structure of the team. These higher level social factors are discussed next.

Organization and Management Problems

Even a group of two, as in a 2-person cockpit, has an authority structure and rules by which work is distributed and coordinated. Yet, the formal structure may not fully describe the interpersonal dynamics involved in actual crew operations. For instance, the clear-cut captain versus first officer role and task distinction may be undermined by dozens of other status indices, such as seniority, training credentials, age, gender, race, and rank (in the military). Airline mergers, retirement procedures (in which captains can choose to fly as second officers rather than retiring), and union disputes all can contribute to creating ambiguity with respect to the authority structure as formally defined. Foushee (1984) described numerous examples of aircraft accidents that have been related to status-induced communication problems. As teams increase in size and tasks become more specialized, other problems relating to multiple and heterogeneous crews are generated (e.g., sub-team management and interface, and the coordination of tasks having potentially conflicting goals).

In space and other isolated, confined environments, a factor that seems to emerge consistently is the "us versus them" phenomenon that occurs between the isolated group and "outsiders." Nicholas, Foushee, and Ulschak (1988) note:

> Visitors to Salyut have attested to being shunned by cosmonauts, and travelers to Antarctic stations have even been refused shelter by research-

> ers. . . . The more a group is isolated, the greater its resistance to outgroups, including the larger organization. . . . Us vs. them has contributed to problems in the past, including the "strike" by astronauts on Skylab IV. (p. 5)

The Skylab missions in 1973 and 1974 were unique in U.S. space history because of their relatively long duration (1, 2, and 3 months). Crews lived and worked in space, often under near mission-abort conditions; yet a great deal of online equipment repair and replacement was accomplished in addition to an agenda of densely scheduled scientific work. Although mission control was clearly a part of the same organization, the resentment apparent during the first half of Skylab IV's mission involved astronauts' feelings toward ground managers. While there are many factors that can explain some of the irritation (e.g., space sickness, faulty equipment, work overload), the degree of tension between this crew and mission control was notable (also see Balbaky, 1980; Connors, Harrison, & Akins, 1985). Anecdotal accounts from observers of the Soviet space program have also noted these kinds of problems (e.g., Bluth, 1981; McNeal & Bluth, 1981).

As a result of these observations, it has been suggested that greater group autonomy may be needed in order to compensate for some of the effects of long-term isolation. Ultimately this is a policy issue, for the authority structure can be modeled in a number of ways (democratic, military), and these decisions will be closely linked to leadership and training philosophy. Whether organizational policy ever results in the desired changes in crew behavior is highly dependent on the specific way in which policy is implemented through training and evaluation programs, and whether adequate interventions or countermeasures are developed. These issues are discussed next.

Training and Intervention Problems

This final problem area is distinctive because it is not only a legitimate research area on its own, but it is also the means by which many of the other problems can be solved. For example, previous sections have focused on the distinction between individual technical skills and crew coordination skills (or cockpit resource management, CRM, as it is commonly called in the aviation industry). One obvious solution to the problem is to incorporate CRM training into the standard programs in which technical skills and procedures are trained and evaluated. Although this strategy has been taken by a number of airlines and military organizations, it is a time-consuming process to incorporate this training philosophy at all levels of the organization (management, instructors, and evaluators in addition to the trainees). It is probably equally obvious that training cannot solve all "crew problems" even if the best training is available to all individuals. There must be legitimate avenues for an individual to dissent from other members of the group, as well as the means to end or control conflict or agreement when it occurs. While training programs have traditionally considered leadership issues and techniques, they have seldom attempted to consider what makes for good "followership" and how the two roles mesh most effectively.

Crisis intervention is a critical aspect of any training program that deals with an environment in which there are high-risk safety factors. Most of us are familiar with the flight attendants' safety briefing to passengers, for example. Yet we know that numerous complications can occur in a real emergency, and we hope that the crew will know how to improvise in case the emergency is *not* of the textbook variety. Is the emergency training really adequate for the real-world problems that can occur?

A recent report on the Chernobyl disaster stated that emergency procedures existed for such an accident, but they were never practiced because such training would scare the workers. If this is an accurate assessment, one must evaluate the costs and benefits of such procedures as they are clearly ineffective if not practiced. In addition to possible psychological "costs" (such as fear), the reluctance to practice emergency procedures may be partially due to inherent risks in the emergency procedure itself. For example, passenger evacuation from an airplane contains some element of risk because people can injure themselves in the deplaning process. While the risks are minor in comparison to consequences of an actual emergency situation (e.g., bomb threat, onboard fire), it may be quite difficult to motivate people to practice activities that can potentially injure them.

In addition to emergencies, the space environment's long duration, isolation, and extreme physiological demands due to microgravity place new stresses on crew members, stresses that may grow to crisis proportions if left unchecked. While the prevention of emergencies occurring in space is preferable, preparation for emergencies is clearly an important area to study thoroughly. A first step might be the development of stress-monitoring instruments and guidelines that would allow early detection of physiological and psychosocial problems.

The problem areas discussed above do not represent a comprehensive listing of critical problems in the aerospace domain; rather, they constitute theoretically distinct areas that are probably relevant for nearly any workplace. It should be evident from the discussion herein that these areas are actually highly interdependent and potentially interact at many levels. The next section of the chapter incorporates them as part of a larger conceptual framework where their interrelationships are further explored. Subsequently, specific examples are given as to how the investigation of these problems may be approached within the constraints of the aerospace environment.

Research Approaches

The tremendous range of problems discussed above necessitates a corresponding breadth in research strategies. However, several overriding concerns should be mentioned. First, research focused on any one of these problem areas should not isolate itself from the system within which it is embedded. For example, it should be obvious that we cannot learn how a team functions by testing solely for individual skills, nor by testing in settings that do not adequately represent the task environment. Similarly, but less obvious, we cannot investigate the way in which individuals relate to each other without adequately representing the *social* environment within which the task is carried out. In short, we must be careful that the research paradigm itself does not control out, or make irrelevant, the phenomena we wish to study. Case studies, in which the results will be directly applied to the situation and group studied, avoid many inferential statistical constraints; yet even there the problem of measuring complex and interacting behavior in an objective, systematic way is no less difficult a task.

Research Philosophy

When conducting basic research, issues of validity become increasingly salient, and the trade-off between experimental control and generalizability is a major design decision. Mindful of practical constraints, we advocate the following minimum requirements: (1) that the social aspects of the situation are adequately and "meaningfully" represented, and (2) that the fidelity of the task situation is appropriate to the research question. If, for example, we were interested in the effects of isolation and confinement on astronauts, it would be questionable, at best, to use typical college sophomore subjects placed in a room with no meaningful task or purpose other than to get class credit. Though motivation is not the primary construct of interest, we must at least be able to assume that there is some comparable purpose or motivation that makes our subjects and the target population (astronauts in this case) analogous. In a flight simulation experiment where we can plan a close match between subjects and target population (e.g., pilots as subjects and a pilot population), as well as between task environments (high-fidelity simulator and aircraft), there is much that can be done in the way of scenario (task) development that makes for the difference between a "batting practice" type of performance and a line-oriented full-mission (LOFT[2]) performance (see Lauber & Foushee, 1981). The "batting practice" analogy applies to the typical pilot training session, wherein pilots are given emergency after emergency with little regard to realism other than the simplest aspects of man-machine interface. LOFT is a training concept that models a flight scenario to an actual mission. It is more selective about abnormal conditions, and it builds in all the complexity that normally would occur in operations. This includes all radio transmissions, company procedures, realistic weather developments, and consideration of traffic in the airspace. From a methodological standpoint, an appropriately described LOFT scenario can provide an excellent opportunity for studying total crew performance.

An Integrated Research Approach

Figure 6.1 displays a variety of research approaches, each of which offers some clear benefits as well as some shortcomings. Two dimensions characterizing these approaches are experimental control (displayed vertically) and fidelity of research environment (on the horizontal axis). Because there are trade-offs associated with each approach, the ideal solution is to maintain an integrated program that can use a

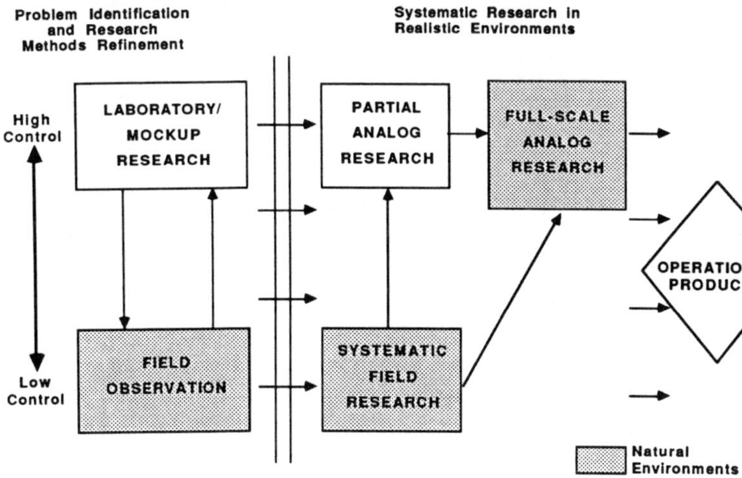

Figure 6.1. An Integrated Research Approach

number of methodologies in order to converge on a particular problem. Each approach is particularly suited to answering certain types of questions, and these answers may be valuable precursors to be fed into another research paradigm. The arrows between boxes in the figure represent linkages where results from one part of an integrated program can benefit another part.

Laboratory/Mockup Research

Experimental control, precision of measurement, and accessibility by researchers are probably the strongest benefits of this type of research, while costs vary greatly with respect to initial installation and maintenance of a working laboratory, mock-up, or test bed. Unfortunately, there are numerous drawbacks due to low fidelity correspondence to the very complex tasks and environments that are typically of interest to applied social psychologists. Nevertheless, laboratory and mock-up research may be used to great advantage in testing smaller, specific part-task questions, for identifying problem areas, and for refining methods and instruments. This may be particularly useful when actual or analogous environments—for space operations, for example—are either inaccessible or excessively costly. For instance, some specific telecommunications issues might be addressable with a fairly low-fidelity workstation mock-up.

Field Observation

While field observation research falls at the opposite end of the spectrum with respect to experimental control, it often fills a function similar to laboratory research—that is, it helps to identify problems and develop research instruments. While this technique typically does not allow for analyses more sophisticated than descriptive statistics, it is, nonetheless, a part of the critical phase of formulating the research questions. Even anecdotal data can be extremely useful in this respect, although they would not contribute in a rigorously inferential way. Accessibility and cost vary greatly, depending on what and where the "field" is (field studies in space are nearly prohibitive, while field studies in airplane cockpits are merely difficult), and there is no question that laboratories offer easier access. Yet, observations made in the field offer unique opportunities to see real-time operations in their full social context.

Systematic Field Research

With a good understanding of the field setting and armed with appropriately tailored research instruments, a researcher can collect a large variety of systematic data in a natural environment in a relatively unobtrusive way. In addition to using paper-and-pencil instruments such as self-reports and observer ratings, other real-time observations can be made, augmented by video and audio tape recordings and interviews. While systematic data collected by these various means will permit generalizability, this is still a setting in which conditions cannot be manipulated as controlled experimental variables. Collecting field data in aircraft operations, for example, has been very productive in investigating a number of topics (e.g., fatigue: see Graeber, Foushee, Gander, & Noga, 1985). However, a great number of variables, such as weather, airspace, and aircraft conditions, will always have uncontrollable yet significant influences on each crew performance. In order to treat such factors as random variation, the level of data analysis must be appropriately independent of the effects of these factors. On the other hand, systematic field research is probably the only experimental setting in which operations can be analyzed within their organizational context. If crews are removed from that context, certain aspects of the work environment will be lost (unless successfully simulated), and crew behavior will lack that element of realism. Thus, a real trade-off is involved: In order to retain high-level organizational fidelity, a certain amount of experimental control may need to be sacrificed.

Partial Analog Research

This methodology is an interesting compromise, as it attempts to maximize the fidelity of the research environment without sacrificing experimental control. High-fidelity aeronautical simulation is a good example of how this methodology can succeed, as the hardware itself is the mechanism by which realism as well as experimental control is maintained. Thoughtful scenario development (such as in LOFT, discussed above) can further increase realism, and appropriate selection of subjects can provide more than just a pilot sample—a flight crew can be provided, along with its company culture and, to some extent, its organizational structure. Space vehicle simulation, such as the shuttle flight deck, can be designed along similar lines. Within the limits of realism, high levels of workload requiring total crew involvement can be attained, as well as the precision of measurement and accessibility that are not possible in a field study. The ease with which this methodology is utilized depends greatly on what workplace is being simulated. What would be a low-fidelity mock-up for a space station might be a partial analog for a factory setting and a full-scale analog for an office setting.

Full-Scale Analog Research

Full-scale analog research builds in a mechanism for control in a natural setting. However, because it is often impossible to validate research conclusions in the actual environment of interest, an analogous environment is used—that is, a natural environment that is both similar in critical respects and accessible to researchers. In the space domain, for example, we might be interested in scientists working in a space station for a relatively long duration, living and working in a confined environment where close contact and coordination with mission control and other ground-based scientists must be maintained. On the basis of limited field observation (in Skylab, for example) and on other partial analog research, we might want to validate a hypothesis regarding crew size and structure. Because we strongly prefer to test this concept prior to an in-space implementation, a functioning undersea habitat that matches the situation in the relevant ways might serve as a reasonable research analog. However, this strategy is still relatively costly and inconvenient in terms of accessibility, and it should be treated as a final phase of validation as opposed to an exploratory testing ground.

On the other hand, analogous environments do not have to support full-scale analog research alone. Productive field observation and systematic field research can be carried out there as well and, in fact, such fieldwork may provide important preliminary data leading to an eventual full-scale experiment. Just as space operations will eventually consist of a variety of workplaces, a variety of analog environments will be appropriate. For instance, in addition to using a partial analog approach (i.e., a simulator) for studying flight deck shuttle operations, actual aeronautical flight operations could be considered a full-scale analog (excluding the long-duration aspect). The future space station, as mentioned above, is nicely paralleled by underwater habitat operations, while teamwork in planetary exploration might use the mountain-climbing environment as an analog. Its harsh and unforgiving setting creates a high-risk situation in which teams must work very closely together to succeed, so it is an ideal setting for considering small group dynamics and leadership issues in stressful conditions. Another harsh, isolated analog that includes the longer duration aspect of space operations is Antarctic winter-over operations, where larger group structure and team interaction issues must be resolved in a stressful, unforgiving environment.

Few research programs can support large-scale projects integrating every kind of approach, and such an integration is not being suggested here. On the other hand, many investigators already informally incorporate these approaches by making field observations and conducting small experimental pretests in the design phase of their primary projects. The message here is that very often a project hits a dead end when pursuing a single paradigm, and there may be ways of stretching that research farther by integrating a different approach, the results of which may complement or converge with other work. We have found that in the aviation domain, a combination of systematic fieldwork with full-mission simulation studies captures both the exploratory and experimental sides of the research. On the space side of our research program, we will probably continue to conduct fieldwork in environments analogous to space operations, but we will want to impose more experimental control into the field settings (i.e., a full-scale analog) in the future. When appropriate, mock-up research can be used to refine some of the more global findings from the field.

Conceptual Framework

In considering research on some subset of the problems discussed above, we will use an adaptation of McGrath's (1964, 1984) input-process-output model as a simple framework for conceptualizing potential factors in a single integrated system (see Figure 6.2). Input factors are those attributes that characterize the individuals making up the group, attributes that characterize the group as a whole, and characteristics of the task environment itself. Output variables are the outcomes of the group – some related to how well the group performed and others that describe how group members felt about the task, their performance, and each other. Process variables are the ways in which the group works together to accomplish their task (i.e., interaction processes, such as communication, problem solving, and decision making).

Input Variables

Input variables are separated into three main categories (all of which may influence one another): individual-level factors, group-level factors, and environmental/task-level factors. In general, they represent relatively static attributes brought to the task situation as opposed to those generated within it. Typical studies of individual-level factors focus on personality, demographic variables, and individual differences in task-related skills that characterize members of a group. Translating these factors to the flight deck, we might consider how a crew is composed with respect to personalities (e.g., Chidester & Foushee, 1988), or investigate the related issue of leadership (e.g., Ginnett, 1987).

Consideration of leadership or managerial style also involves group-related factors, such as group structure, hierarchy of authority, group size, history, and cohesiveness. When teams either interface with other

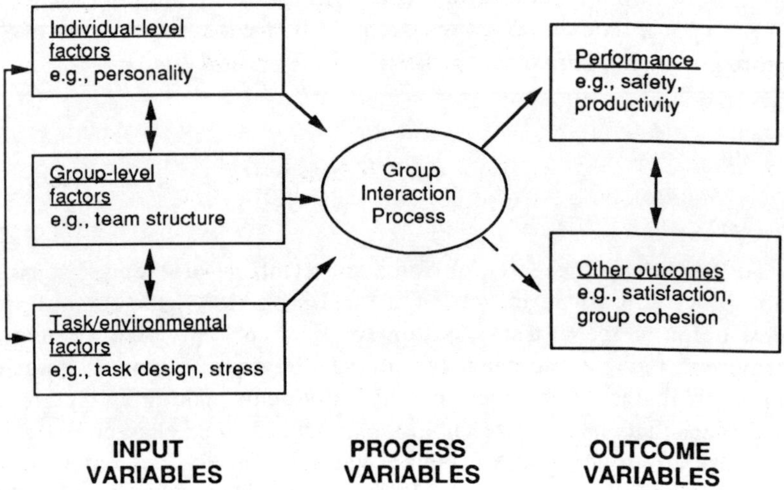

Figure 6.2. Conceptual Framework
SOURCE: Adapted from McGrath (1964, 1984).

teams or are made up of specialized subteams, the organizational structure may be very complex and fluid. The degree to which interfacing teams work autonomously or in concert with one another, and the way they make the transition from one state to the other, may significantly influence the quality of the final performance. Counterproductive tensions between interfacing teams sometimes emerge because the goals of the teams are competitive or conflicting, thereby leading to only partial success for the team as a whole. Similarly, problems can arise between interfacing teams because the authority structure is not clear or is perceived to be inappropriate. For example, astronaut teams may feel that mission control should not have the authority over some decisions because it does not have the information or experience to fully appreciate the astronauts' situation. Control problems of this sort can lead to the often-reported "us versus them" hostility.

Environmental/task input variables include concrete variations, such as the arrangement of instruments or sequencing of checklist procedures; they may also refer to features of a task that are more difficult to measure, such as the degree of workload or stress, and the status and reward structure of the task. In the space domain, the importance of studying environmental factors is great because they affect overall habitability in addition to specific task performance; therefore, they include such elements as privacy, noise level, environmental control, and numerous other factors related to eating, sleeping, relaxing, general health, and hygiene. In both aeronautics and space, the replacement of manual functions with automated ones has motivated a great deal of attention to human-computer interaction and its effect on task performance. Implications of this research are equally important for initial task design and for later training for the end-user.

Outcome Variables

Outcomes in a task setting are usually defined by fairly distinct performance goals, which can be measured in terms of product quality, quantity (productivity), or time (efficiency). However, it is obvious that trade-offs, such as errors, safety problems, and employee dissatisfaction, may also enter into the performance formula. On the flight deck, for instance, "performance" may be defined in terms of on-time departures and arrivals; yet, this simple set of goals presupposes that numerous and complex federal aviation regulations are upheld, that safety is not jeopardized, passengers are not abused, and a certain amount of courtesy is shown to others via various communication channels. Other outcomes, such as the cohesiveness of the flight deck crew, may promote a productive attitude during task performance but may not have wider relevance outside the workplace.

In long-duration spaceflight, however, the importance of task performance side effects, such as team satisfaction or cohesiveness, may be greatly magnified because of the isolated and confined environment. Such outcomes may significantly interact with concurrent performance measures, as well as feed back into the process (much like input factors) to affect later team effectiveness.

Process Variables

Group process variables are the means by which input factors influence outcomes (see Hackman & Morris, 1975; Hackman, 1986;

Lanzetta & Roby, 1960). For example, a relationship may be found between different leadership strategies (input factor) and levels of team performance (outcome), but the means by which leaders produce these effects may be through variations in communications styles (a group process variable). Similarly, differences in checklist procedures (a task input factor) may be related to making safety-related errors (outcome), but the process that drives this relationship may involve differences in decision-making processes that are either promoted or hindered by the procedures themselves.

Probably the most critical feature of process variables is that they represent a dynamic and interacting sequence of actions that occur in real time—not a simple individual-difference measure. The patterns that emerge are interactively determined. That is, they comprise an individual's actions conditional on the actions of the other(s) in the group.

Relationships Between Variables: Selected Studies

From a conceptual perspective, a full three-part test of the above model would be interesting; from a methodological perspective, however, this is rarely possible. Furthermore, there is no theoretical reason to expect that there should be a simple one-to-one correspondence across all three variables (e.g., that the same process variations will relate to both input and outcome factors identically). Rather, it is more feasible to test the relationship between two variables at a time, most commonly the input-outcome relationship. A natural follow-up to such a study would be to explore a process interaction that relates to one and/or the other.

The remainder of this chapter is devoted to a more detailed description of two aerospace studies, each of which represents a different research phase and methodology. The first is an aeronautical simulation study of the effect of leader personality on aircrew effectiveness, and the second is a systematic field study in the working undersea habitat known as Aquarius.

Study 1: Leader Personality

The first study was motivated by the crew and interpersonal problem area, and it focused on the effects of leader personality on crew effectiveness. (For a detailed description, see Chidester, Kanki, Foushee, Dickinson, & Bowles, in press.) It can be represented conceptually as a study of an input-outcome relationship, and it utilized the partial analog research paradigm. Benefits of this approach include a high degree of experimental control, low risk, and easy accessibility. Short-duration simulations, such as this one, are not good analogs for long-duration spaceflight, but they do provide a partial analog for shorter duration flight operations, such as shuttle pilot operations.

Twenty-three 3-person crews participated in this high-fidelity full-mission simulation study of a Boeing 727 flight deck. Subjects were all professional pilots from a single airline, and the task scenarios were designed to model typical routes, airspace conditions, and weather these pilots would commonly encounter. The simulation was designed to resemble a 2-day duty cycle as closely as possible, including briefings; the normal flight documentation; closely simulated interaction with ground, maintenance, air traffic controllers, and the company dispatcher. Over the two days, crews flew five flight segments, three the first day and two the second. On the third and fifth segments, abnormalities were programmed into the scenario that required that the crews engage in problem-solving crew coordination. The third segment began normally, but a combination of system failures resulted in a runaway stabilizer that jammed in a nose-down condition. Shortly thereafter, an engine low-oil-pressure warning light illuminated, thus creating a relatively high work load situation, but one which was not further complicated by weather or traffic problems. The fifth segment involved deteriorating weather conditions that forced crews to miss the approach (landing was prohibited due to poor visibility) and find an alternate airport. During the missed approach, a hydraulic system failure occurred, thereby increasing aircraft handling difficulty as well as requiring a switch to manual operation of flaps and landing gear. While neither the jammed stabilizer (on the first day) nor the hydraulic system failure (on the second day) are extreme problems, they were sufficiently complicated that a lack of timely planning and coordination could build into a dangerous situation.

Table 6.1
Defining Characteristics of Captains' Personality Clusters

		Characteristics	
Symbol	*Personality Cluster*	*High Level*	*Low Level*
IE+	Positive Instrumental/Expressive	instrumentality achievement striving expressivity	negative instrumentality negative expressivity
I–	Negative Instrumental	negative instrumentality verbal aggressiveness competitiveness	expressivity
Ec–	Negative Expressive	negative expressivity	instrumentality achievement striving

Crews were composed on the basis of the captain's personality profile (Chidester, 1986) as determined by an individual-differences questionnaire largely based on the Personality Attributes Questionnaire, or PAQ (Spence & Helmreich, 1978). Three profiles were identified by performing a cluster analysis on the questionnaire data, which combined the captains into three groups based upon their similarity to one another along a specified set of dimensions (see Anderberg, 1973). The personalities of the other two members of the crew (first and second officers) were left to vary randomly. The captains were characterized as: (1) both highly instrumental and expressive, IE+, (2) negatively instrumental, I–, or (3) negatively expressive, Ec– (see Table 6.1). Captains in the IE+ cluster can be described as high in task and achievement striving, as well as high in interpersonal skills; thus, theoretically, they are the most likely to exhibit good leadership qualities. Captains in the I– cluster are also very task-oriented, but their arrogance and competitiveness result in a more individual achievement motivation, as in the "right stuff" pilot. Captains in the Ec– cluster tend to express themselves in a negative way (complainers) and are characterized by relatively lower levels of instrumentality and achievement striving (although pilots, as a group, are high in both instrumentality and expressivity compared to the normal population – see Chidester, 1986).

Crew outcomes were defined by performance measures based on several different sources: ratings made on-line by an expert observer from within the simulation cab, and a compilation of crew errors as

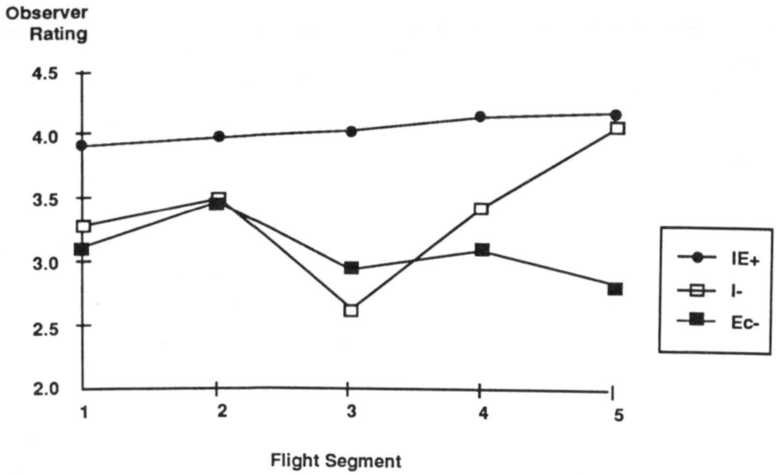

Figure 6.3. Crew Composition Simulation Findings on Leader Personality and Crew Performance
SOURCE: Reprinted from Chidester, Kanki, Foushee, Dickinson, & Bowles (in press).

judged by observers of the videotaped segments. Aircraft handling data were also available directly from the simulator computer system for detecting technical flight deviations.

The relationship found between captain personality and crew performance can be seen in Figure 6.3. Note especially the differences in flight segments 3 (Day 1), and 5 (Day 2), as these were the ones in which abnormal operations occurred. On the basis of performance defined by observer ratings, crews led by a captain fitting the positive instrumental and expressive profile (IE+) showed consistently effective performance across all flight segments. Negative expressive (Ec–) captains led consistently less effective crews across all flight segments. Interestingly, negative instrumental (I–) captains led crews that were less effective on the first day, but highly effective on the second day (particularly on abnormal flight segment 5). In essence, the improvement in performance suggests a familiarity effect; that is, crew members have "learned" to coordinate better through experience. Unfortunately, we cannot know from this analysis alone whether the captains modified their behavior or whether the subordinate crew members simply learned to accommodate to the "right stuff" demeanor (as is often reported by junior crew members in real life). However, we will be able to address these questions in our later analyses of crew communication, because far greater detail can be observed from the videotape records of each of these aircrews.

Given these interesting input-outcome relationships, an analysis of the group process characterizing these crews is a natural next step. Many questions might be asked, but our initial strategy consists of an exploratory investigation of communication patterns that can be found to differentiate effectively performing crews versus less effective crews, and/or the three crew compositions as defined by the captains' personality profiles. The analyses will be based on detailed transcriptions of all verbal exchanges and significant flight events during a normal and an abnormal flight segment on Day 1, as well as a normal and an abnormal flight segment on Day 2. From the transcriptions, speech events will be coded and arranged in real-time interactive sequences in order to discover distinctive patterns. The eventual goal will be to identify specific communication processes relating to leader personality differences and (in a separate analysis) ones relating to variations in crew performance. For an example of a study based on a different simulation situation, but using the above strategy for investigating the relationship between communication process and performance (the process-outcome link), see Kanki, Lozito, and Foushee (1989). This study found speech patterns that distinguished high-error performances from low-error performances of 2-person aircrews in a high-fidelity simulation study conducted by Foushee, Lauber, Baetge, and Acomb (1986). Specifically, question-answer and command-acknowledgment sequences appeared to be used differently by high-error versus low-error crews. The overriding finding was that low-error crews showed marked homogeneity in their speech patterns, thus indicating the use of some standard or "conventionalized" form of communicating, while high-error crews showed a great diversity of patterns, thereby indicating lack of an established convention.

Study 2: Undersea Habitat

The second study to be described here is a currently continuing systematic field study utilizing the undersea habitat, Aquarius, as an environment analogous to space station operations. Although field research cannot typically support a traditional experimental design with great control and precision of measurement, part of the research strategy is directed toward that end—that is, toward eventually designing an effective full-scale analog study. In the meantime, the field setting allows maximum generalizability (high external validity) and provides the opportunity to investigate numerous problem areas simultaneously. In conceptual terms, we are able to explore multiple relationships between input, outcome, and process variables.

The undersea habitat Aquarius is located in the Salt River Canyon, St. Croix, U.S. Virgin Islands, and is operated by the National Oceanic and Atmospheric Administration (NOAA) in conjunction with Fairleigh Dickinson University. The facility is made up of three components: a place (wet porch) where divers enter and leave the habitat, an entrance lock that connects the other two components, and the main lock (a tube that is 21 feet long having a 9-foot diameter) that constitutes the entire indoor living/work space for a crew of 6 (5 science crew members and 1 operations engineer). A typical mission is 10 to 14 days, including a 16-hour decompression procedure, which allows a habitat team to surface safely after having been saturated[3] at the habitat depth of 50 feet, or 2.5 atmospheres (see Zorpette, 1988, for an interesting description of Aquarius). The habitat is an attractive research setting because the saturated environment allows relatively long underwater time not possible given the limitations of surface divers.

Aquanauts are usually teams of marine scientists. Their activities are closely integrated with a topside operations team whose job is to maintain the facilities and safety backup, and to provide training, equipment, and other operational resources. The communications and coordination characterizing the science/operations interface is a particularly good analogy to the astronaut/"mission control" interface described previously. When the science team arrives (usually including topside science support members in addition to the aquanaut team), several days are reserved for training prior to the first day of saturation. At this time, the operations team briefs the science team on a variety of topics: geography of the area, dive navigation, procedures for communicating and planning activities, medical issues, and emergency procedures both within the habitat and during dives. All teams work together to get equipment and supplies ready to be taken down to the habitat. Once aquanauts are settled in the habitat, the operations team provides its resources and is responsible for their safety. Specifically, the team maintains a 24-hour watch desk from topside that monitors the life support system and habitat by means of multiple audio and video sources. In addition, no aquanaut ever dives outside the habitat without a safety boat team keeping watch from the surface. The topside science support team also provides resources to the aquanauts (e.g., research supplies, information), but it typically works through the operations team. A laboratory space is allocated to this support team, and often there is daily lab work that is coordinated with the aquanauts' data collection. Thus, there are three distinct teams working together during a mission: the aquanaut team (composed of up to 5 science team members and 1 habitat engineer), the topside operations team, and the topside science support team.

At the crew and interpersonal level, the science team, under the leadership of a principal investigator (PI), must coordinate its efforts toward achieving its science goals in this totally life-supported environment. How crew composition variables and leadership styles relate to crew performance is an important topic, although the definition of "performance" is an issue in itself. Because habitability issues intensify the need for compatibility in the workplace, it is not clear whether productivity alone can be deemed a good measure without considering other outcomes, such as adherence to safety and other regulations, crew member satisfaction, and reactions to stress and tension. As Hackman and Morris (1975) have noted, "devices for improving group effectiveness fall into two classes: (a) interpersonal techniques . . . intended to improve the quality of the relationships among group members; and (b) procedure-oriented techniques, which provide group members . . . with specific strategies for going about their work on the task in a more effective manner" (pp. 92-93). In the habitat environment, an argument can be made for considering both types of intervention techniques and outcomes.

Task and environmental factors are obviously important in this regard because they impose significant constraints on behavior. How dives are scheduled, whether teams engage in shift work, and how, in general, work is distributed—all these factors are relatively free to vary in this workplace. The environment, however, imposes significant constraints: for instance, the duration of a dive, the frequency of dives within a time period, and (as a safety precaution) the number of divers that must work together.

General habitability issues, such as preparing meals, eating, housekeeping, and sleeping, can also be investigated in this setting. For example, does it matter that these activities probably depart from one's typical routines? Similarly, will the constraints due to space; lack of privacy; isolation from others; changes in noise level, air, and temperature in the habitat; and the overall physiological changes due to living in a pressure of 2.5 atmospheres eventually cause problems that interfere with task productivity?

The NASA/NOAA Undersea Habitat Project began in 1988, and data from 10 science missions have been collected thus far. Because a number of problem areas are being addressed and many of the critical variables are still in the process of being defined, data collection is being carried out in as comprehensive and diversified a way as is feasible without infringing on the mission operations. During the training days prior to saturation, a NASA observer gathers three types of self-report data from the aquanaut team members: general demographic background and experience questions, a measure of individual differences (the PAQ, identical to the one used in Study 1), and baseline information about food and sleep practices and preferences. These data all reflect input factors—attributes of the people and group that are brought to the task situation. In addition, data regarding social/organization factors are obtained from participants' curricula vitae, official organizational charts, and observer notes. Data regarding the task itself are provided by a science briefing given by the PI and by the written science proposal. Information about the habitat and underwater environment is provided by operations manuals and interviews with operations staff.

Outcome variables in this setting can be summarized in three main categories: science productivity, operational safety, and crew member satisfaction. While it is not yet clear whether these aspects of performance always interact, there are clear cases where a good science day can generate overall team satisfaction. There is also evidence that safety concerns can sometimes undermine science productivity and that interpersonal tension can sometimes undermine team coordination. Thus, the operational definition of performance is still an open question, although we are concerned with systematically integrating the three aspects listed above.

Performance data are collected in a variety of ways. On a daily basis, each aquanaut is asked to give his or her assessment of the activities that took place. This includes science progress, teamwork, satisfaction, and questions relating to habitability issues. These self-report data are accumulated during the saturation and turned in at the end of the dive.

The observer monitors activity and communications from topside by taking daily notes of departures from the normal and obtaining information from the operations dive log, an important document that accounts for all planned and actual dive activities. (Similar to an airline's departures and arrivals, deviations from the planned schedule can result in extra work, inefficient performances, and aggravation.)

At the end of the mission, aquanauts are given a final post-dive assessment questionnaire in which they evaluate the entire time they functioned as a team in the habitat living/workplace. In addition to productivity, team coordination, and individual satisfaction with the mission, topics included bear directly on environmental issues, particularly food and sleep. Crew members are also asked to rate one another on several dimensions in order to gain sociometric information about the group. Again, outside evaluations are provided by the observer and a member of the operations staff; together they provide a broader perspective on how this team performed in comparison to others. The purpose of multiple evaluations is to develop a more integrated definition of performance (i.e., one that incorporates many aspects of performance from a variety of perspectives). This in turn will allow for a direct test of its relationship to a large number of input factors that have been identified.

As critical input variables are identified and an integrated outcome variable is defined, a methodology for capturing group process is also being developed. Currently, two approaches are being explored, both of which utilize videotape samples of behavior taking place in the main lock of the habitat. Audio input is supplied by speaker microphones in the main lock, direct input from the sound-powered telephone (which links the main lock to both operational support staff at the watch desk and their own science support team in the topside lab), and all VHF radio transmissions on a dedicated NOAA frequency. Given the multiple audio input, speech can sound chaotic at times, both within the habitat and at the topside watch desk. However, the extra channels provide flexibility and necessary redundancy.

Interaction processes within the aquanaut team will be investigated by analyzing communication data from videotapes of a structured group activity. From prior observations, it has become clear that teams vary greatly in the way they structure their activities; hence we could not count on a naturally occurring activity that would involve all members in the main lock *and* that would be comparable for all teams. Therefore, group tasks (team-building games) were introduced to thc group as a recreation activity that could be taped consistently for all teams.

Because we also wanted to focus on the interaction between the topside operations team and the aquanaut team, daily videotapings were made, starting an hour prior to the first dive of the day (according to the schedule). If the dive did not conform to the intended schedule, reasons for deviation were noted by the observer. The dive event is important because it requires that the aquanaut team's plan be well scheduled, that the PI coordinate his/her crew members, and that topside operations and topside science support be given ample notice for preparing diving equipment and science supplies, respectively. In addition, topside operations and science support teams must coordinate their activities as well. When teams are extremely rushed or some members are waiting excessively for others, operations can suffer in terms of both productivity and interpersonal relations. Because data are in the process of being collected and an analytical strategy is still being developed for some of the habitat measures, findings cannot yet be reported. However, it is clear from our informal observations that the several problem areas discussed earlier in this chapter are well represented.

Conclusion: Operational Products

The studies described above exemplify a particular research philosophy and an integrated research approach. They bear on several problem areas and represent numerous input-process-outcome relationships. This final section describes the types of operational products that are sought by the aerospace community/industry. While directly related to the problem areas, such operational products represent the applied results of research, the *technologies* that can be designed on the basis of the lessons learned from research.

While we would not expect to generate operational products in the individual/physiological domain, the importance of incorporating these factors into our research design cannot be overlooked. In essence, these types of factors represent the limiting conditions on the crews and must be systematically controlled (or manipulated) in any consideration of human performance.

In the crew and interpersonal domain, issues of selection are critical. We need crew selection technologies suited for various sizes and types of crews (e.g., flight deck versus space station) that select not only for individuals but for team compositions that will be most effective. Once teams are formed, an obvious starting point for inducing behavioral improvements in crew interactions is to design a communication assessment instrument that identifies effective communication strategies and provides a means of training for these skills.

Results from studies of environmental and task issues are theoretically most effective if they are implemented at the design stage—for instance, in designing tasks that integrate human and automated components, or in constructing work/rest schedules near an environment with altered day/night cycles. However, some environment and task factors are both fixed and stressful and, in these cases, research can aid in the development of countermeasures for reducing stress related to the task environment. Countermeasures to physiological stress due to microgravity and to the more psychological stresses associated with living in an isolated, confined environment are of obvious importance in future space missions.

Research in the appropriate analogous environment can yield information that is highly useful in developing guidelines for an optimum organizational structure. Specifically, when parameters such as crew size, mission duration, and task goals are known, it will be desirable to have a means for modeling a system that is maximally efficient in terms of authority and information transfer. However, organizational structure cannot in itself guarantee that principal participants will operate in a way that takes advantage of the predetermined structure. A training program will need to accompany the structural design in order to promote effective leadership and interaction both within and between teams.

In a sense, the last problem area, training and intervention, includes types of operational products that overlap with nearly every other problem area. Insofar as we are principally interested in input factors and group processes that relate to performance, training and intervention techniques are critical in (1) monitoring crew productivity and well-being, and (2) providing methods for ameliorating problems or crises that occur, whether these be matters of crew coordination, social/organizational structure, or factors relating to the task environment itself.

In aerospace research, we are working not only in a physical environment in which there are many unknowns, but in a social environment that is dynamic and changing. Humans introduce their own unique dynamic qualities to an environment; they cannot simply be viewed as organisms that respond in systematic ways, but rather as organisms that can creatively change the meanings and relationships that we, as researchers try so painstakingly to delineate.

Notes

1. ASRS compiles a confidential data base of reports from pilots, air traffic controllers, and other members of the aviation community and is jointly operated by the National Aeronautics and Space Administration and the Federal Aviation Administration. It is a validating source of information for delineating problem areas.

2. Line-oriented flight training is a technique in which crews fly a complete trip in a high-fidelity flight simulator. Flights are made in real time, and no intervention is made by the instructor regardless of crew actions. It is a learning experience in which an emergency situation is usually included, and the errors made become the topic of a fairly comprehensive debriefing following the performance.

3. As a function of depth and time, a diver's bloodstream becomes saturated with dissolved inert gases, such as nitrogen. Although a return to the surface in a saturated state could be lethal due to the expansion of gases that block the bloodstream, divers may remain in a saturated state by living in the habitat (at a depth of 50 feet, from which they can work at greater depths and for longer time periods). Prior to their return to the surface, they "decompress" within the habitat over 16 hours as pressure is slowly lowered to the surface level and excess nitrogen is freed from the bloodstream.

References

Anderberg, M. R. (1973). *Cluster analysis for applications.* New York: Academic Press.

Balbaky, M. (1980). *Strike in space.* Cambridge, MA: Harvard University Case Clearing House.

Baron, S. (1988). Pilot control. In E. L. Wiener & D. C. Nagel (Eds.), *Human factors in aviation* (pp. 347-385). New York: Academic Press.

Bluth, B. J. (1981). Soviet space stress. *Science, 81,* 30-35.

Chidester, T. R. (1986). *Mood, sleep, and fatigue effects in flight operations.* Unpublished doctoral dissertation, University of Texas at Austin.

Chidester, T. R., & Foushee, H. C. (1988). Leader personality and crew effectiveness: Factors influencing performance in full-mission air transport simulation. *Proceedings of the 66th Meeting of the Aerospace Medical Panel on Human Behavior in High Stress Situations in Aerospace Operations* (Conference Publication AGARD-CP-458, pp. 7/1-7/9). Neuilly sur Seine, France: Advisory Group for Aerospace Research and Development.

Chidester, T. R., Kanki, B. G., Foushee, H. C., Dickinson, C. L., & Bowles, S. V. (in press). *Personality factors in flight operations: I. Leader characteristics and crew performance in full-mission air transport simulation* (NASA Technical Memorandum 100728). Moffett Field, CA: NASA-Ames Research Center.

Connors, M. M., Harrison, A. A., & Akins, F. R. (1985). *Living aloft: Human requirements for extended spaceflight* (NASA SP438). Washington, DC: NASA Scientific and Technical Information Branch.

Cooper, G. E., White, M. D., & Lauber, J. K. (Eds.). (1979). *Resource management on the flight deck* (NASA Conference Publication 2120). Moffett Field, CA: NASA-Ames Research Center.

Curry, R. E. (1985, May). *The introduction of new cockpit technology: A human factors study* (NASA Technical Memorandum 86659). Moffett Field, CA: NASA-Ames Research Center.

Dolgin, D. L., & Gibb, G. D. (1988). *A review of personality measurement in aircrew selection* (NAMRL Monograph No. 36). Pensacola, FL: Naval Aerospace Medical Research Laboratory.

Ellis, A., & Conrad, H. S. (1948). The validity of personality inventories in military practice. *Psychological Bulletin, 45*, 385-426.

Foushee, H. C. (1984). Dyads and triads at 35,000 feet: Factors affecting group process and aircrew performance. *American Psychologist, 39*, 885-893.

Foushee, H. C., & Helmreich, R. L. (1988). Group interaction and flightcrew performance. In E. L. Wiener & D. C. Nagel (Eds.), *Human factors in aviation* (pp. 189-227). New York: Academic Press.

Foushee, H. C., Lauber, J. K., Baetge, M. M., & Acomb, D. B. (1986, August). *Crew performance as a function of exposure to high-density short-haul duty cycles* (NASA Technical Memorandum 88322). Moffett Field, CA: NASA-Ames Research Center.

Foushee, H. C., & Manos, K. L. (1981, September). Information transfer within the cockpit: Problems in intracockpit communications. In C. E. Billings & E. S. Cheaney (Eds.), *Information transfer problems in the aviation system* (NASA Report No. TP-1875). Moffett Field, CA: NASA-Ames Research Center.

Ginnett, R. G. (1987). The formation of airline flight crews. In R. S. Jensen (Ed.), *Proceedings of the Fourth International Symposium on Aviation Psychology* (pp. 399-405). Columbus: Ohio State University, Department of Aviation.

Graeber, R. C., Foushee, H. C., Gander, P. H., & Noga, G. W. (1985). Circadian rhythmicity and fatigue in flight operations. *Journal of University of Occupational and Environmental Health, 7* (supplement), 122-129.

Hackman, J. R. (1986). *Group level issues in the design and training of cockpit crews* (NASA Conference Publication 2455, pp. 23-39). Moffett Field, CA: NASA-Ames Research Center.

Hackman, J. R., & Morris, C. G. (1975). Group tasks, group interaction process, and group performance effectiveness: A review and proposed integration. In L. Berkowitz (Ed.), *Advances in experimental social psychology* (Vol. 8, pp. 45-99). New York: Academic Press.

International Civil Aviation Organization. (1984). *Accident prevention manual* (1st ed.). ICAO Doc. 9422-AN/923.

Kanki, B. G., Lozito, S. C., & Foushee, H. C. (1989). Communication indexes of crew coordination. *Aviation, Space, and Environmental Medicine, 60*, 56-60.

Lanzetta, J. T., & Roby, T. B. (1960). The relationship between certain group process variables and group problem-solving efficiency. *Journal of Social Psychology, 52*, 135-148.

Lauber, J. K., & Foushee, H. C. (1981). *Guidelines for line-oriented flight training* (Vol. 1, NASA Conference Publication 2184). Moffett Field, CA: NASA-Ames Research Center.

Lautman, L. G., & Gallimore, P. L. (1987, April-June). Control of the crew caused accident: Results of a 12-operator survey. *Airliner: Magazine of the Boeing Commercial Airplane Co.*, pp. 1-6.

McGrath, J. E. (1964). *Social psychology: A brief introduction*. New York: Holt.

McGrath, J. E. (1984). *Groups: Interaction and performance*. Englewood Cliffs, NJ: Prentice-Hall.

McNeal, S. R., & Bluth, B. J. (1981, May). *Influential factors of negative effects in the isolated and confined environment*. Paper presented at the Fifth Princeton/AIAA/SSI Conference on Space Manufacturing, Princeton, NJ.

National Transportation Safety Board. (1972). *Aircraft accident report: Eastern Air Lines, Inc. L-1011, N310EA, Miami, Florida, December 29, 1972* (Report No. NTSB-AAR-73-14). Washington: Author.

Nicholas, J. M., Foushee, H. C., & Ulschak, F. L. (1988). Crew productivity issues in long-duration space flight. In *Proceedings of the American Institute of Aeronautics & Astronautics 26th Aerospace Sciences Meeting* (Conference Publication AIAA-88-0444). Washington, DC: American Institute of Aeronautics and Astronautics.

Ruffell Smith, H. P. (1979). *A simulator study of the interaction of pilot workload with errors, vigilance, and decisions* (NASA Technical Memorandum 78482). Moffett Field, CA: NASA-Ames Research Center.

Sexton, G. A. (1988). Cockpit-crew systems design and integration. In E. L. Wiener & D. C. Nagel (Eds.), *Human factors in aviation* (pp. 495-526). New York: Academic Press.

Spence, J. T., & Helmreich, R. L. (1978). *Masculinity and femininity: Their psychological dimensions, correlates, and antecedents*. Austin: University of Texas Press.

Wiener, E. L. (1985). Beyond the sterile cockpit. *Human Factors, 27*, 75-90.

Wiener, E. L. (1988). Cockpit automation. In E. L. Wiener & D. C. Nagel (Eds.), *Human factors in aviation* (pp. 433-461). New York: Academic Press.

Wiener, E. L., & Curry, R. E. (1980). Flight deck automation: Promises and problems. *Ergonomics, 23*, 955-1011.

Wolfe, T. (1979). *The right stuff*. New York: Farrar, Straus, & Giroux.

Zorpette, G. (1988, March). An underwater visit to Aquarius. *IEEE Spectrum: Magazine of the Institute of Electrical and Electronics Engineers*, pp. 24-29.

7

Human-Machine Interactions in Spaceflight

HARVEY WICHMAN

Behavioral scientists have long had an interest in the responses of individuals and groups in settings where the extreme nature of the environment was the independent variable. Scientists have studied the behavior of prisoners in concentration camps (e.g., Vaughan, 1947), crewpersons in undersea habitats (e.g., Miller, VanDerwalker, & Waller, 1971), crewpersons at South Pole stations (e.g., Mullin & Connery, 1959), submarine crews (e.g., Kinsey, 1959), and astronauts in U.S. and Soviet space stations (e.g., Banks & Ride, 1989; Compton & Benson, 1983). In an article on the psychosocial factors affecting simulated and actual space missions, Kanas (1985) provides a good brief review of the above types of studies. In most cases, these studies were nonexperimental and therefore lack rigorous controls. However, behavioral scientists have often been involved in the design of an environment—or of procedures to be carried out in that environment—and selection of the persons who would perform in the environment. As a consequence, quasi-experimental studies have been possible, in which results obtained in the environment have indicated the effectiveness of the original designs, plans, or selections.

In spite of all that has been written about behavior in extreme environments, as the 20th century draws to a close, we still have very little solid knowledge to give us confidence in planning for life in orbiting space stations, long-term interplanetary trips in spaceships, or colonies on moons or planets. This chapter describes some research that I have been involved with in the U.S. space program, and then discusses some of the ways in which behavioral scientists can participate more fully in the type of research that will lead to successful applications of psychology in the space program.

Something New in Social Psychology

Social psychology is a difficult field to define because it "is actually a hybrid discipline that inherits problems, data, and concepts from both parents [psychology and sociology]" (English & English, 1958, p. 509). One area of interest that might not normally be included within social psychology is the interaction of persons with machines. However, because of new developments in technology, a situation has now developed in which people and some machines can interact socially. Thus, social psychologists are the scientists with the relevant skills, attitudes, and knowledge to deal with some of the issues that are now beginning to appear, especially in the computer sciences and some of their applications to the space program.

In Arthur Clarke's story, *2001, a Space Odyssey* (and the popular movie by the same name), there was a very sophisticated computer called the *HAL 2000* aboard the interplanetary spacecraft (Clarke, 1968). The computer was able to respond to spoken commands and reply by means of a quiet and pleasant-sounding synthesized voice. Of course, the computer could be accessed and respond in the traditional ways as well. The spacecraft crew members all referred to the computer as "Hal" when speaking or referring to it. The social interactions between the crew and the computer became a central aspect of the story, and this *social* interaction between machinery and humans was experienced as eerie or even frightening by many who saw the movie.

When the movie was released in 1968, a computer that could carry on a conversation was logically possible and hence allowed for a plausible story. But the reality of such a device was then so remote that it could only occur, with any sense of credibility, in a science fiction movie. At the time this book went to press, the equivalent of the *Hal 2000* computer is still a very long way off. However, computers that can be spoken to and can in turn reply with electronic voices now exist and are beginning to find serious practical applications in medical technology, aviation, and a host of other commercial areas. There are numerous ways in which this technology could be effectively employed in space operations, such as onboard the future U.S. space station *Freedom* (see Wichman, 1987a). One such space application is described below.

In the current generation of space suits, information regarding the status of a suit's various systems is sensed electronically, stored in computer memory, and shown to the astronaut on a display located on the chest pack. This information can be accessed by pressing a button which, with each successive press, serially presents the data for each system. An irritant for the astronauts is that if the button is held down for too long, the unit scrolls past some information; the astronauts then must go through the entire list to get back to the display of the desired information. Manipulating buttons is very difficult for astronauts because their fingers are encased in inflated gloves, thereby resulting in poor tactile sensitivity. Generally, the astronauts interrogate their suit subsystems about once every 30 minutes. Because getting suited up to go outside a space vehicle is a lengthy and difficult task, extra-vehicular activity (EVA) is usually planned for 6-hour periods (about three-fourths of the duration of the life-support systems). During an EVA, a pair of astronauts will each have to stop working in order to interrogate their suit about a dozen times. This clearly is a wasteful and frustrating use of an astronaut's time. (One estimate of the value of an astronaut's EVA time in 1986 dollars was $21,000 per hour – Wichman, 1987b!)

Almost certainly, the next generation of space suits will employ a speech recognition/speech synthesis system to allow astronauts to communicate by voice with their suits (Wichman, 1987b). During a time when no other communications are taking place, an astronaut wearing such a suit could continue with a routine task while at the same time asking the suit to announce the status of the various systems. The suits' electronic voice units would then speak the necessary information into the headphones of the respective astronauts.

Experiences with Voice Recognition

During the year I spent as a Sloan Foundation Fellow with a team at Rockwell International that was doing preliminary design work on the space station, I worked with various voice recognition systems in the Rockwell laboratory. Anyone who has ever worked with computers knows that they can be very frustrating at times. However, I seemed to notice that I became more irritated with computer errors when I was entering information by voice than when typing in the information. As I observed other scientists working in the lab, the same thing seemed to be true of them. I then had the opportunity at Edwards Air Force Base to observe test pilots flying a General Dynamics F-16 fighter plane with an experimental "cockpit of the future." This cockpit had a voice computer system that allowed the pilots to accomplish a number of tasks by voice, such as tuning the radios by speaking to the computer instead of having to look down into the cockpit and turn the knobs manually. For instance, a conversation might go something like this:

Pilot: Tune radio. Comm one. One two niner decimal seven.

Computer: [Spoken in the pilot's headphones by an electronic female voice] Comm one. One two niner decimal seven.

Pilot: Execute.

Computer: [Automatically dials the correct frequency into the number-one communication radio and announces] Task completed.

The system was programmed so that commonly used radio frequencies could also be referred to by words. For instance, the pilots could simply ask the system to dial in "Edwards Tower," and the system would dial the correct frequency for the control tower at Edwards Air Force Base. While skimming across the desert close to the ground at high speed and in very bright sunlight, it was advantageous to not have to remove a hand from the controls and look down into a dark cockpit to change a radio frequency manually. But when I asked the test pilots how they *felt* when mistakes occurred following a verbal entry of data—as opposed to manual entry—they agreed that they seemed to get more irritated following voice entry.

Clearly, an experiment was called for that would determine whether this difference in emotional responses was in fact occurring and, if so, would then find some way to quantify it. We conducted such an experiment (Wichman & Polson, 1988), but before it is described below, readers need an understanding of a state-of-the-art voice interactive system. Without that understanding, it would be hard to imagine how interacting with a computer could produce the kind of psychological responses discussed herein.

Because a demonstration cannot literally be given in a book, the next best option is a simulation, embodied in a script of an interaction scenario between a person and a computer. The following scenario is a valid example of what can be done with personal computers at the present time. While the scenario has been modified for the purposes of this chapter, most of the features described exist in a system currently in operation at the Jet Propulsion Laboratory in Pasadena, California. The system was designed and developed by Wichman and Lynch (1988).

Simulated Computer Interaction Script

Operator: [Turns the computer on]

Computer: I am on-line. To whom am I speaking?

Operator: Ann Wichman.

Computer: Stand by, please.

[The computer contains the voice templates for the names of each user so that it will recognize his or her name upon replying. There is a pause while the computer loads the entire dictionary of Ann's voice prints into its working memory. This represents all of the words that the computer would recognize, if they are spoken by Ann. Ann will have prerecorded them earlier. Each user of the system will also have to prerecord his or her words.]

Computer: Good afternoon, Ann. What can I do for you?

Operator: I want a menu.

Computer: Okay. Here is the list of menus.

[The computer knows the time of day and makes the appropriate greeting. It would have given a specific menu if requested. Because only a menu in general was requested, the artificial intelligence algorithm assumes she needs the list of all the menus and gives it to her. Here is what appears on the computer monitor.]

LIST OF MENUS AVAILABLE
1. Hydraulic System
2. Electrical System
3. Life Support System
4. Water System
5. Waste System
6. Pressure System

Operator: Give me the electrical system, please.

Computer: Thank you. Here it is.

[Here is what next appears on the monitor.]

ELECTRICAL SYSTEM
1. Battery System
2. Solar Cell System
3. Solar Dynamic System
4. AC System
5. DC System
6. Voltage Color Code

[The computer has a tit-for-tat strategy built into its algorithm that encourages people to be polite. If the computer is spoken to in an impolite fashion, it will respond in kind. The following is an illustration of what would happen if Ann were rude.]

Operator: Give me a damn menu!

Computer: You got it, foulmouth!

[Continuing now, Ann makes another request.]

Operator: Give me the voltage color codes, please.

Computer: Thank you. Here it is.

[Here is what next appears on the computer screen.]

VOLTAGE	COLOR CODES
1. 0-6	Blue
2. 7-14	Green
3. 15-28	Yellow
4. 29-110	Orange

Operator: Give me a cursor.

Computer: Okay. Here it is.

[At this point, a cursor appears in the upper left corner of the computer screen and remains there, blinking on and off.]

Operator: Position the cursor.

Computer: Will do.

Operator: Down, down, down, down, down, stop. Type.

[The computer moves the cursor down one line following each command until it is stopped one line below the number 4 on the screen. The word "type" instructs the computer to shift to the typing mode and to listen for numbers, letters, and punctuation. It is very easy to confuse short words and letters, so the operator must speak in the International Phonetic Alphabet. Ann wants to add a fifth line to the list with a new voltage range, and give it the color code red. As she says each item, the computer will respond by typing it on the screen until it has written, "5: 111-220 Red." Notice below what happens when she inadvertently spells red incorrectly:]

Operator: Five, period, space, one, one, one, tab, dash, space, two, two, zero, tab, tab, Romeo, Echo, Delta, Delta, Whoops! – uhmmm, backspace.

[Here the computer responds by moving back a space and erasing the extra "d" from "redd." Exclamations such as "whoops" and "uhm" are very difficult to deal with if they sound like some other word in the set of voice templates. Some algorithms are designed to recognize and ignore them, or to respond to them in a cute way such as having the computer say "goofed up again, eh?"]

Operator: Return. Save.

[The new line of text is now in place and permanently saved. The next instruction will shift the computer out of the typing mode.]

Operator: Cancel cursor.

[Ann can see on the screen what is happening, so no verbal reply is needed. Therefore, there is none, but Ann wants to check to see that the computer understands and wants it to give her a verbal reply. Therefore, she so informs the computer.]

Operator: Reply verbally. What color is the 220-volt color code?

Computer: It is red.

Operator: I have a new request.

Computer: Go ahead.

Operator: Telephone call.

Computer: Okay. Who should I call?

Operator: Harvey Wichman.

[The computer checks its memory for the phone number of Harvey Wichman and finds that it has three numbers for him. Therefore it asks:]

Computer: Should I call him at his home, office, or laboratory?

Operator: At his laboratory, please.

Computer: Okay. Please stand by. I am placing the call now.

[The computer selects the proper phone number from its memory and places the call using its modem.]

Harvey: [Answers his ringing telephone] Hello?

Computer: [Over the phone line to Harvey Wichman] One moment, please, I have a call from Ann Wichman.

Computer: [To Ann over the computer's loudspeaker] Ann, your party is on the line.

Operator: Thank you. Go to sleep now.

[It is necessary to inactivate the computer so that upon hearing Ann speak to Harvey, the computer doesn't think she is speaking to it. We use the phrase "go to sleep," which shifts the computer to a memory bank that has only one phrase in it: "Wake up." Until it hears that phrase, it remains essentially dormant. The phrase shifts the computer back to its main memory of Ann's word templates. The reason for choosing the phrase, "go to sleep," is that it is a phrase not likely to occur by accident in an office setting where it would unexpectedly put one's computer out of commission.]

Computer: Okay, I am going to sleep. Bye-bye for now.

[So that the operator cannot forget what state the computer is in, this message appears on the screen.]

STATE OF COMPUTER
ZZZZZZZZZZZZZZZzzzzzzzzz

Operator: [Picks up her telephone and speaks into it.] Hello?

Simulated Script (Uninterrupted)

The following unannotated version of the same interaction script will impart a better sense of the flow of the interaction.

Ann: [Turns the computer on]

Computer: I am on-line. To whom am I speaking?

Ann: Ann Wichman.

Computer: Stand by, please. [Pause] Good afternoon, Ann. What can I do for you?

Ann: I want a menu.

Computer: Okay. Here is the list of menus.

Ann: Give me the electrical system, please.

Computer: Thank you. Here it is.

Ann: Give me a cursor.

Computer: Okay. Here it is.

Ann: Position the cursor.

Computer: Will do.

Ann: Down, down, down, down, down, stop. Type. Five, period, space, one, one, one, tab, dash, space, two, two, zero, tab, tab, Romeo, Echo, Delta, Delta, Whoops! – uhmmm, backspace. Return. Save. Cancel cursor.

Ann: Reply verbally. What color is the 220-volt color code?

Computer: It is red.

Ann: I have a new request.

Computer: Go ahead.

Ann: Telephone call.

Computer: Okay. Who should I call?

Ann: Harvey Wichman.

Computer: Should I call him at his home, office, or laboratory?

Ann: At his laboratory, please.

Computer: Okay. Please stand by. I am placing the call now.

Harvey: [Answers his ringing telephone] Hello?

Computer: One moment, please, I have a call from Ann Wichman. Ann, your party is on the line.

Ann: Thank you. Go to sleep now.

Computer: Okay, I am going to sleep. Bye-bye for now.

Ann: [Picks up her phone] Hello?

Reading this scripted scenario should give a sense of the repartee that can take place between a person and a computer even with the primitive systems that exist at this time. Let us now return to discussing evidence from our experiment (Wichman & Polson, 1988).

An Experiment with Voice Recognition

Kristen Polson and I drew up the preliminary design of an experiment in which we would program a computer to commit certain errors (unbeknownst to the participants) following both voice entry and manual entry. We then discussed the design with researchers in the field of voice technology, most of whom concluded that we would find greater negative emotional responses following voice entry. The researchers believed that persons *talking* to computers would assume that if an error occurred, it would be the fault of the computer for misinterpreting the instruction. However, if persons were *typing* information into computers and an error occurred, the participants would attribute the fault to their own typing mistakes. They would therefore get angry at the computer in the former case because they thought the error was the computer's fault. In the latter case, where they attributed the fault to themselves, they would not get as angry as if the fault were attributed to the computer.

Notice how these people, who regularly interacted by voice with computers, had begun to speak about them in terms of attributions and misattributions of blame—social psychological terms and principles that are usually restricted to human interactions (or occasionally to interactions with other living creatures, such as pets).

Thus it became clear that we had to design our experiment so that no doubt existed regarding attributions on the part of the participants. It had to be clear to everyone that the error was definitely a malfunction of the apparatus. Then, if participants still got angrier following voice inputs than when following the use of manual inputs, the cause could logically be attributed to the act of speaking to the machine.

Design and Procedure

Participants sat at a personal computer that was programmed to present 60 randomly chosen nine-digit numbers, one at a time. All participants saw the same numbers. After a number was presented on the computer screen (without a computer voice), the participant was to match it either by typing or by speaking the correct digits into a lightweight boom microphone with a head clamp. Each participant typed in 30 numbers and voice input 30 numbers, and all 60 were displayed in the same way on the screen—one line below the stimulus, thus facilitating checking the two for correspondence before entering the response digits. The participants were randomly assigned to one of two groups, which experienced the typing and speaking conditions in reversed order to control for order effects.

Unbeknown to the participants, the computer had been programmed to give an error message four times during manual entry and four times during voice entry (always in the same location: numbers 10, 18, 25, and 28). After having entered the correct matching number, the participant was informed by a message presented on the screen that the number was in error and should be reentered. Thereupon the screen was erased, the same number was presented by the computer, and the participant was to match it once again. After the second successful match, the computer accepted the number and went on to the next one.

In order to determine the emotional response of the participants to the machine's errors, we measured two aspects of emotionality: its quality and quantity. The qualitative aspect was determined by observing the participants' overt responses to the error (e.g., blurting out, "Hey, this thing is stupid") and by interviewing the participants afterwards. The quantitative aspect of the emotional response (the participants' level of arousal) was assessed by measuring the change in heart rate that occurred in association with the error announcement. Each participant wore a finger plethysmograph pickup on the index finger of the nondominant hand. Pulse rate readings were taken each time a participant pressed or said "return" in order to go on to the next item. Following the error message, heart rate was monitored continuously until a participant pressed or said "return" after rematching the numbers; because heart rate was somewhat erratic during this period, the peak heart rate was used as the measure of emotional intensity.

Participants in this experiment, consisting of 21 students recruited from the five Claremont Colleges, were tested individually. They were seated in front of an IBM-XT personal computer that contained an Introvoice VI speech recognition unit. After hearing the instructions, they were given a chance to ask questions, and the equipment was connected. Then they had to "train" the computer to recognize their voices by speaking into the computer, according to a programmed system of prompts, all of the words and numbers the computer would have to recognize (10 times for each word and number). The entire experiment took each participant approximately 50 minutes.

Results

The mean heart rate for all participants following nonerror entries was 72 bpm (beats per minute). While this is the national norm for the resting heart rate, it is somewhat high for this age group, and the elevation probably was caused by the vigilance associated with this activity as well as the slight general arousal a stranger often gets from the novelty of the laboratory. For the two seconds following the error message, there was usually a decrease in heart rate, which was then followed by a dramatic increase. This is typical of the orienting reflex, which occurs when an organism experiences a novel stimulus (e.g., the unexpected error message). During the orienting response, heart rate is diminished and breathing halts, making the whole body a very stable sensing platform (Grossman, 1967); its most dramatic manifestation is the freezing of activity seen in a wild animal that has just heard an unexpected sound. The dependent measure of arousal was the peak heart rate that occurred after the orienting response, which took place while the participants were reentering the matching number after having been informed of the error. The mean increase in peak heart rate following an error in the manual entry condition was 1.3 bpm; in contrast, in the voice entry condition, it was 5.8 bpm—a very significant difference [t (20) = 3.69, $p < .001$]. It is worth noting that, overall, voice entry was about as effective as manual entry in that the unprogrammed error rates for each were nearly identical. In addition, by inspection, the data indicated that there were no fatigue or order effects.

The qualitative aspect of the emotional response was very clear to an observer. Participants would become agitated, frowning and fidgeting in their seats. Comments included, "this machine is cheating me," or "this thing is weird." In the manual entry condition, the reentry of the number was typically done with slightly more powerful keystrokes. However, in the voice entry condition, the reentry was often done with an "angry voice" that produced a voice wave-form that did not match the original voice template that had been spoken in a "calm voice." Upon hearing such an angry voice the computer would then print on the screen, "Say again," thereby compounding the frustration of the participant.

Discussion

Earlier I proposed the hypothesis that people would show greater negative emotional responses to machine errors following voice input than they would following manual input (Wichman, 1987b). The data in this study support that hypothesis and quantify the effect: The mean peak heart rate increase following an error during voice entry was 4.4 times greater than that for manual entry.

There is a deeper level of analysis that these data do not address, however: Why is this happening? We are currently undertaking studies to answer that question, but we have some anecdotal evidence suggesting what the answer might be. From observing the participants and interviewing them immediately after their involvement in the experiment, it seemed that they made numerous anthropomorphic references to the machine when making voice entries that they did not make when doing manual entries. When they would mutter under their breaths following an error, there seemed to be more personifications of the machine during voice entry than during manual entry. Thus it seems that when people operate machines by speaking to them, they come to personify them.

This, of course, is still an untested hypothesis. There are other possible explanations for the phenomenon that need to be considered. One idea, which comes from the perceptual defense literature (e.g., McGinnies, 1949), is that the act of saying something that has emotional loading produces an amplification effect. The lore of romance points out that terminating a love relationship is more easily done in a letter than by the spoken word. Psychotherapists often use the technique of having clients write about something that they are too anxious to discuss. Perhaps, then, the mild arousal that occurred following a manual entry error (1.3 bpm) was simply amplified (5.8 bpm) by having spoken the entry. The mystery is still unresolved.

Planning for the New Space Station

Human factors psychology, a branch of applied psychology, has a long history of dealing with person-machine interfacing. However, this discipline has traditionally dealt with structural and functional aspects of the person-machine interface, not aspects that are social and emotional. Such issues as the readability of dials, the most desirable shape of knobs to minimize user confusion, and the time lag between control inputs and machine responses are characteristic of human factors work. In fact, a synonym for human factors is *ergonomics*, a term that grew out of early human factors research on variables that produced fatigue in the operators of mechanical equipment.

Now, however, advances in technology have led us into an era in which new person-machine concerns are arising, some of which would have been hard to imagine only a few years ago. As so often happens, empirical scholars were preceded in this area by creative writers. In retrospect, Arthur C. Clarke's (1968) account of the social interactions between the *HAL 2000* computer and its operators was impressive for its richness and psychological sophistication. It is now incumbent upon empirical social psychologists to employ the tools of science to explore realistic aspects of human-machine interactions in space flight.

Some of the psychological issues arising during space flight can be studied by analyzing the experience of crews in the U.S. space shuttle flights. However, it is not safe to use the experience of shuttle crew members to make inferences regarding working conditions on the future space station. Only rarely do shuttle flights last as long as 10 days—a week or less is more typical—and there are still very few crew members who have been on more than one or two flights. Space flights of short duration are such exhilarating experiences that crew members can tolerate all sorts of inconveniences. In contrast, crews on the space station will probably be in orbit for 90-day periods.

This difference in time spent in flight is a very important reason why the shuttle is not a good analogy for the space station, but there are some others as well. The shuttle orbiter cabin is much more restrictive than the working and living spaces on the space station. The kinds of tasks undertaken in the orbiter cabin (which is essentially a cockpit on the upper deck) are different from those that will take place in the space station laboratories. Another key difference is that if anything goes seriously wrong or an astronaut becomes ill, the shuttle crew can return immediately to earth. Presently, there is no way to return space station crew members except to send a shuttle orbiter to get them, a process that could take as long as 28 days! In regard to crew evacuation, astronauts consider a top-priority need to be a "space lifeboat" that would be docked to the space station and allow crew members to return to earth at their will. If such a vehicle is not produced, the sense of isolation and loss of control will be much greater in space station crews than in shuttle crews. I do not mean to say that the shuttle cannot be used for research relevant to the design of the space station; however, it is important to note that there are several shortcomings inherent in trying to generalize from the shuttle to the space station.

The Soviets, of course, have extensive space station experience, and a considerable amount of that information is now generally available (e.g., Bluth & Helppie, 1986). In addition, the United States had a space station called Skylab that was occupied by three crews in 1973 and 1974, and we have a considerable amount of valuable information from those experiences (e.g., Compton & Benson, 1983). Nevertheless, there is much yet to be learned if we are to develop a well-designed space station and deploy it in the mid-1990s.

Spacelab

The space shuttle can be used as a better analogy of the space station when it carries aloft the Spacelab, a cylindrical laboratory module designed to be carried in the cargo bay of the shuttle orbiter (Shapland & Rycroft, 1984). It is about 7 meters (23 feet) long and about 4 meters (13 feet) in diameter. While Spacelab is only about half as long as the modules planned for the space station, it is nearly as large in diameter; therefore, it is a good representation of the type of work space that will be available in the laboratories and resource nodes of the space station. Once the decision was made that the shuttle would be the primary means for the United States to put materials for a station into orbit, the shape of the modules for the space station was determined: whatever goes up will have to fit into the cargo bay of the shuttle and thus must be in the shape of a large cylinder.

The tasks performed in Spacelab are similar to those that will be performed in the space station. Spacelab allows crews to work in the high-level vacuum and micro-g of low earth orbit. In a typical 7- to 9-day mission, Spacelab is secured inside the orbiter's cargo bay. The crew rides into orbit in the orbiter's cabin and then enters the Spacelab via a 1 m-diameter tunnel that connects with the air lock in the orbiter mid-deck area. The temperature, air pressure, and humidity are essentially the same in the orbiter cabin as in the Spacelab module, providing a shirtsleeve working environment for the crew. Crew members who aren't on duty spend their time in the orbiter cabin, where everyone also eats and sleeps.

Spacelab has flown twice—once in November and December 1983 for 10 days aboard the orbiter *Columbia*, and again in April and May 1985 aboard *Challenger* for eight days. Spacelab experiments on the two flights were conducted in materials processing, environmental observation, fluid mechanics, astronomy, and the life sciences. Figures 7.1 and 7.2 show views of the Spacelab, including the orientation of crew members at workstations.

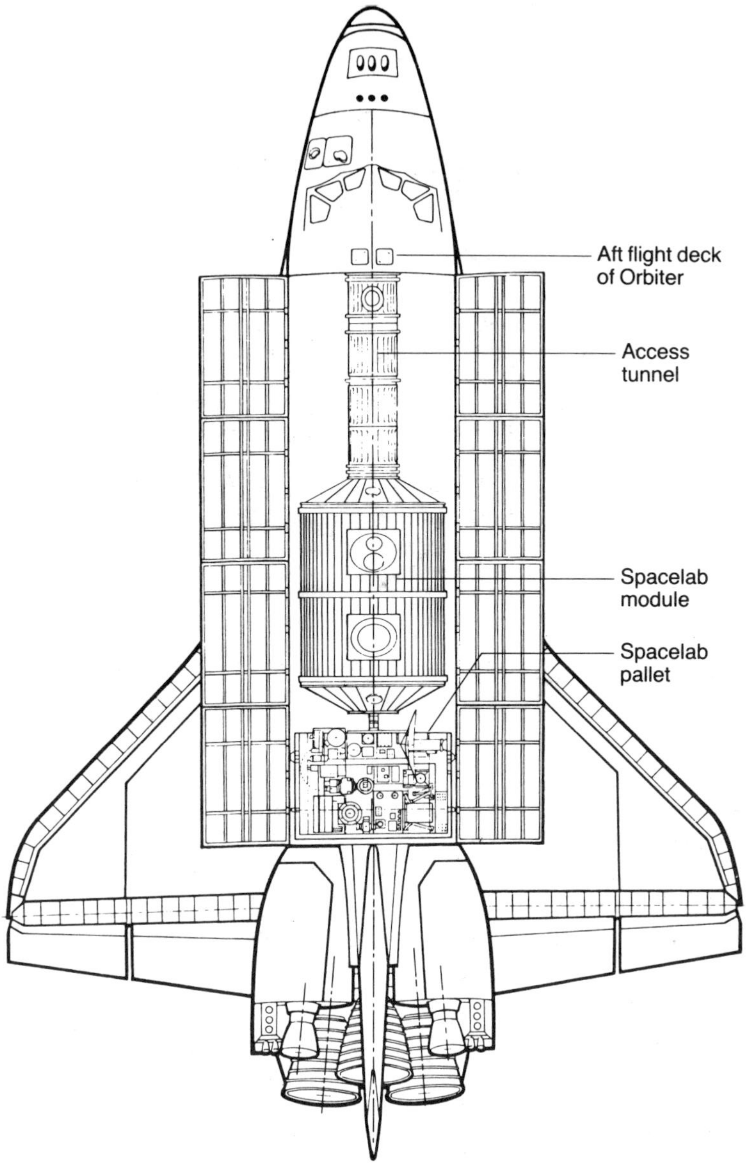

Figure 7.1. Position of Spacelab in the Shuttle's Cargo Bay
SOURCE: From Shapland & Rycroft (1984).

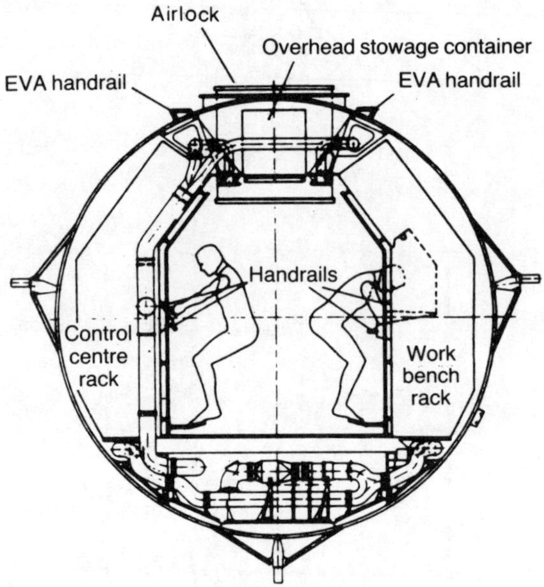

Figure 7.2. Crew Working Positions in Spacelab
SOURCE: From Shapland & Rycroft (1984).

Spacelab provided a very good opportunity for behavioral scientists to study the behavior of crew members at work in a weightless condition inside a space vehicle. Many other aspects of space flight, such as isolation, crowding, and danger, can be simulated on earth with moderately high fidelity. However, except for brief periods of about 40 seconds during airplane flight in a parabolic arc (and except for underwater activity while wearing a space suit weighted to zero buoyancy), weightlessness simply cannot be simulated on earth. Shirtsleeve weightlessness—that is, without the bulky and awkward life-support suit required for extravehicular activity (EVA) in space—as well as its interactions with other aspects of space flight, must be studied in orbiting vehicles such as the Spacelab aboard a shuttle orbiter.

Research on Weightless Behavior

During the two Spacelab flights, a video camera located at the end of the module, opposite the tunnel entrance, gave controllers and scientists on earth an opportunity to see what was going on inside the chamber. More than 50 hours of recordings were made of the video images transmitted back to earth. Of those recordings, about 10½ hours show useful images of crew members at work. Rockwell International's Space Division in Downey, California, made a set of the tapes showing crew behavior for our lab to study.

Watching the tapes quickly established that there were some human factors variables worth studying and quantifying. For instance, crew members seemed to move around a lot from one workstation to another—much more than seemed efficient. In addition, there were frequent periods in which a weightless crew member would drift away from a position of restraint and begin to struggle reflexively (most people who see these episodes find them very funny). Another thing we noticed was that rather than use a pair of cloth foot-loop restraints located on the floor in front of a rack of equipment, the astronauts would grasp a handhold with one hand while working with the other.

The findings of our Spacelab research teams illustrate the kind of human factors research that can be conducted quite easily and inexpensively using videotapes of crew members performing in actual work settings. However, note that these findings come from ex post facto studies that were not planned prior to the flights. The videotapes only show haphazardly collected episodes of behavior because randomly selected time samples of an entire flight were not obtainable. In terms of representative research design, this may be a serious shortcoming. With only a little careful preparation, better research could be conducted. For example, if an entire flight were videotaped, random samples of the entire flight or exact specified portions could be taken. If a second camera were added at the opposite end of the module, any loss of view resulting when one crew member blocks another could be avoided. Minor improvements in lighting and the ability for human factors researchers to operate the cameras by remote control from the ground in real time (e.g., zooming in on a specific behavior) would allow for very sophisticated research. In spite of the limitations of the data, we currently have available views of spontaneous weightless behavior over a lengthy period of time; because that is all we have to work with at this time, it would be a shame not to extract all the useful information that is possible.

Findings

We found that crew members spent about 33% of their time in the Spacelab module on the two flights working with one hand while holding on with the other. Obviously, this way of working is very different from behavior on earth, except in the instance when one is standing on an unstable platform, such as a ship's deck in rough seas. About 33% of the time, crew members used the foot restraints mounted on the floor in front of various workstations. About 10% of the time was spent moving from one workplace to another. Because one cannot walk in a zero-g (gravity) environment, any movement from one location to another is termed *translating*. The crew members seldom spent more than a couple of minutes at any one location, and they seemed to be in motion almost constantly. In most earthbound workstations, such constant motion would be considered very wasteful. Somewhat more than 4% of the crew's time was spent struggling as a result of the loss of adequate restraint. Clearly there is much room for improvement here. The above figures accounted for about 80% of the crew's time in the module. The remainder of their time was either spent suspended without restraint or was classified in an *other* category that did not fit any of our operational definitions.

The Spacelab was designed with a floor, ceiling, and two walls, just like rooms on earth. Technically speaking, there is no need for such a design in space. Because one does not walk in space, and cannot place objects down on a floor, floors are not needed. In zero-g, things do not fall from ceilings (as they do on earth) nor stay put on the floor—they drift off in the air currents. It is clear from the videotapes we saw that the astronauts were operating in an environment designed for use on earth, and often they simply had to do the best they could when in orbit. There were some good reasons for this, however. In order to have a successful 7- to 8-day flight, in which the largest possible number of experiments were performed, crew members trained for more than a year on earth. It was vital for this training to take place in a simulator that exactly duplicated the interior of the space environment. However,

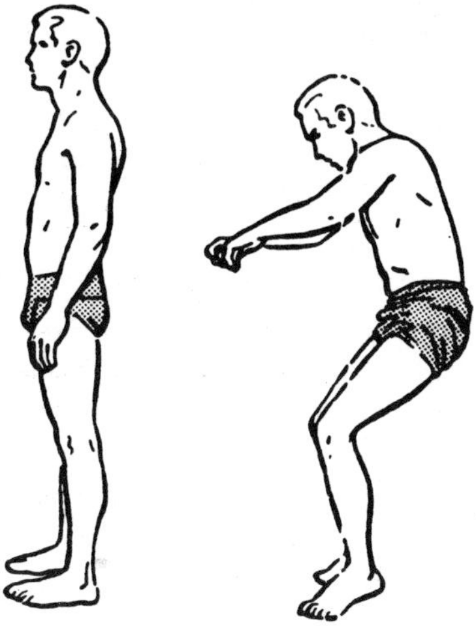

Figure 7.3. Normal Body Posture on Earth and in Weightless Flight
SOURCE: Adapted from Thornton, Hoffler, & Rummel (1977).

astronauts must walk to and fro on earth, and they must stand erect in their normal earth posture as they manipulate levers and knobs. By contrast, in zero-g, the relaxed human body is not erect. The natural tension of muscles and ligaments pulls the body into approximately the position it takes when one is floating upright in water. Figure 7.3 shows an anthropometric drawing of a weightless human in what is known as the *neutral body position*.

In Figure 7.3, notice, for instance, that the sole of the foot is not at a right angle to the shinbone, as it would be in a person relaxed but standing on earth. A crew member in orbit cannot use a simple foot-loop type of foot restraint without constantly expending effort to keep his or her heel close to the floor. The design of foot restraints is a good illustration of how difficult it is to create an adequate zero-g environment for an organism that has evolved for life in 1-g. Some designers have suggested using Velcro on the soles of shoes and on the floor. However, the sturdy Velcro used on earth to fasten garments is not acceptable for space flight because it gives off a toxic gas when it burns. The Velcro used in space vehicles is adequate for holding light objects, such as pencils, but it would not provide enough restraint for the forces applied by a person moving levers or pulling open a drawer.

NASA's constant frustration is that it must train personnel very extensively in a shirtsleeve environment on earth that is quite different from what will be experienced in space. Therefore, the space environment has been designed as a partial compromise. The sooner and more frequently human factors specialists are involved in these designs, the better will be the results. To do their work well, however, these specialists need more information of the type we have been extracting from the Spacelab videotapes.

NASA, obviously concerned about crew restraint in Spacelab, had foot-loop restraints placed on the floor of Spacelab I. Two loops were placed side by side on the floor with their centers about 40 cm (18 in) apart. Though these foot restraints were sparingly used, we found that their use did increase as the flight progressed. This finding suggests that the use of the restraints may not seem natural to crew members at first but that they do learn to use them. In Spacelab II, some foot restraints were placed in a staggered fashion – next to one another but with one slightly in front of the other. We found that the foot restraints were used more often on Spacelab II than on the first flight. Though the foot loops were arranged in pairs, crew members typically used only one with their dominant foot, placing the other foot behind them or wedged against some part of the structure to provide better front-back (longitudinal axis) support. Clearly, this indicated a problem with the side-by-side foot loops; they provided adequate lateral stability but inadequate longitudinal stability. The staggered foot loops corrected for that problem, which may account for their increased use in the second Spacelab flight.

Nevertheless, though the foot restraints were there, the crew members learned to use them, and the staggered foot loops in Spacelab II seemed to represent an improvement over those in Spacelab I, the restraints were not used for long periods of time. The mean time spent with one foot in a restraint was 65 seconds and the mean time spent with both feet in a restraint simultaneously was only 46 seconds. It was obvious that the use of foot restraints as they were currently designed was not very popular with crew members. When not using a foot restraint, crew members were either translating, drifting, or restraining themselves by hand.

One reason for the short durations of foot restraint was that the crew members moved about frequently. This was because many of their tasks were of a short duration, preceded and followed by translation from one equipment rack to another. This need for frequent movement was partly due to the fact that each of the experiments on board the Spacelab was developed independently of the others. The structure of the racks was standard, but there was no central console for control or data gathering. Therefore, much time and effort was spent moving from one experiment to another in order to perform brief control or observation functions. Such a setup is a basic management problem. On earth it would be unthinkable, for example, to design an aircraft cockpit in which crew members had to constantly get out of their seats and move to another part of the cockpit to manipulate a lever or read a dial. Instead, consoles are provided for the pilot, copilot, and flight engineer, with all systems designed to be operated or monitored from some combination of those three positions.

The above information illustrates a problem with using the Spacelab as an analog for the space station. With short-term experiments designed for flights of a week or less, Spacelab's inefficient human factors design was acceptable, but would not be acceptable for 90-day periods. One danger is that the space station may be designed for use in much the same way as the Spacelab (it, too, will have standardized racks). However, human factors data such as ours make it clear now—well in advance—that the space station should be designed for more central monitoring and control than has been the case with Spacelab. It is important to note that even though the Spacelab is not an exact analog for the space station, it is quite easy to project human factors findings from Spacelab to the space station and anticipate where problems will occur.

My hope is that the work we have done using videotape analyses will cause NASA to make human factors research a part of each future Spacelab flight, in addition to the other research taking place onboard. This could be accomplished very inexpensively with a minimum of interruption. It would likely reap important benefits for future space endeavors, as well as for NASA's analogs back on earth regarding undersea habitats and polar stations.

Other Psychosocial Issues in Space Flight

Having discussed two relatively new issues in social psychology—planning for weightless behavior, and the social interaction of people with machines—I now want to address some other social psychological issues that are more traditional, such as training, group dynamics and management, and cross-cultural relationships. These issues neither arise from new technology (such as the ability to speak to machines) nor from new experiences (such as weightlessness) made possible by new technology. Rather, they are enduring issues that must be dealt with anew in the different settings that result from the new technology.

Training

Having grown up in a railroading family, I learned early that crew coordination among engineers, firemen, brakemen, and conductors was of the utmost importance. Likewise, coordination is currently a lively area of research in aviation psychology (e.g., Foushee, 1982, 1984). Just as there are similarities and differences in the problems of crew coordination between train crews and air transport crews, so there will be differences and similarities in crew coordination problems between air crews and space crews. It will be important for psychologists to clarify where the similarities end and the differences begin, and to study the differences as new research areas. This all becomes even more important in relation to the current controversy over whether space exploration should use human crews or be automated. In discussing the value of utilizing humans in space exploration, Loftus (1986) declared:

> This is really the most significant point—people are so much easier to program or reprogram than any system of comparable capability; and the more skilled the people, the easier the reprogramming, much of which can be done by the crew itself in real time. (p. A77)

Making training as effective as possible has become crucial in the space program because so much time is devoted to it. As mentioned earlier, a shuttle crew may train for more than a year in preparation for a one-week flight, and at this time in the history of space travel, it is rare for an astronaut or cosmonaut to have more than three flights in a career. The training demands for even a short-duration flight are so heavy that a common response in my conversations with astronauts and mission specialists has been, "Don't add something new that has to be learned!" Not only must material be learned, but routines must be practiced continually after they are learned in order to assure that skill levels are high at the time of the flight. This is necessary because, once the flight has begun, considerable time may elapse between the last training and the time when the skill must be employed.

A dramatic example is the case of a shuttle commander who will land the vehicle for his first time. The commander will have practiced for many hours in a conventional electronic flight simulator designed to simulate the shuttle; up until just before the shuttle launch, he will practice landings in a Gulfstream II executive twin-jet aircraft that has been highly modified to be a high-fidelity simulator of the shuttle during its landing approach and the flare before touchdown. This Gulfstream II is referred to as the shuttle training aircraft (STA).

In the cockpit, there are two sets of controls. On the left side (the side occupied by the shuttle commander) is a duplicate of the shuttle commander's station, and on the right (the copilot's side) is a set of normal Gulfstream II flight controls. STA training only deals with the small part of the shuttle flight associated with the landing—the letdown from 35,000 feet to touchdown.

> Shuttle pilots all fly the STA regularly and with increasing frequency as their launch date approaches. During the final months before launch, shuttle pilots are in the STA at least once a week, and then fly daily in the final week before launch. As you may expect, trainees are graded most harshly on their first approach of every STA mission, because only the first approach counts on a shuttle mission. The shuttle pilots fly several practice approaches in the STA the day before liftoff, and then they sit in space for several days before flying an "airplane" again. (McClellan, 1985, p. 38)

This example of the extensive training required for only a few minutes of performance at the very end of a space flight illustrates why space crews feel that the training burdens of their occupation are getting out of hand.

As a way of dealing with the possibility that there could be as much as a 28-day delay in getting a rescue mission up to the space station, one NASA proposal was to train at least two crew members as surgeons, capable of dealing with surgical demands as complex as appendectomies. However, most astronauts found this to be a totally unacceptable use of their time, and many were not interested in developing such skills. They have suggested instead that NASA develop a vehicle for emergency return to earth from the space station. Thus it is clear that the idea of any new type of training is not likely to be greeted with enthusiasm by space crews.

Nevertheless, Nicholas (1989) has proposed that training in interpersonal and group-behavior skills be provided for crews on the space station, while Kanas (1988) has suggested that physicians serving as space station crew members be given special psychosocial training. Regardless of what the training implications are, Santy (1987) has pointed out that the psychiatric component of the space station's health-maintenance facility has to be clearly defined and planned.

There are already some rules for space activity that are designed simply to avoid conflict (in lieu of training crew members to deal effectively with conflict). For instance, cosmonauts do not play chess on the Soviet space stations—a rule established because a Soviet crew member once killed another in a dispute over a chess game at a South Polar station. Knowing the psychological hazards of living in severe environments, even for the most highly selected persons, the Soviets decided it was safest to forbid chess playing onboard their space stations.

Any behavioral scientists who are going to work in the space program should understand that the technical training demands associated with operating space flight apparatus are so great that any additional training demands will be rejected unless their utility is fully justified by solid empirical evidence. Even then, behavioral scientists may find themselves in tough competition for time with technical demands, in what is rapidly becoming a "zero-sum time game." Until now, the operators of the space program have managed behavioral concerns primarily by selection, by the use of a simple military-type management system, and by reaping the benefits of an enormous Hawthorne effect, which is a consequence of the special status attached to being a member of a space flight crew. However, as the space program matures, selection procedures will become less stringent, being a member of a space crew will be less special and, with increases in crew size and the involvement of agencies other than NASA, management will become more complex.

Group Dynamics

Heretofore, all extended space flights have involved only very small groups, either dyads or triads. The three U.S. Skylab crews (in 1973 to 1974) were all triads, and most Soviet crews have been dyads, which have had occasional visitors. Furthermore, the Skylab crews were all comprised of military men, one of whom was designated the commander. What we know about behavior in space comes from experience with the simplest (i.e., smallest) groups, and ones having strongly established ranking systems in which even the most basic types of restructuring (such as coalition formation) were either impossible or highly unlikely. Television news clips have shown the human interactions during space flights as being very businesslike but with an aura of camaraderie, and for the most part that has been true. However, there have been some problems that ought to alert us to future potential difficulties.

Events from two of the Skylab missions clearly illustrate problems that can occur. Because the two primary groups involved in a space station mission are the ground controllers and the flight crew, good relations between these two groups are obviously of the utmost importance. Because the flight crew is in orbit, the two groups can only communicate by radio. Their discussions, therefore, are public. As would be true for most people, not being able to discuss matters in private turned out to be a very serious problem for the crew of the third Skylab mission. To understand what took place during that mission, it is necessary to know about some of the events that occurred during the second mission. When the second Skylab crew docked its Apollo module to Skylab and prepared to enter the space station, all three crew members became nauseated with motion sickness. It was several days before all of them had recovered, and by that time the crew was far behind schedule. Clearly, this had been a socially (as well as physically) uncomfortable experience for this crew. Their way of coping was to become a dynamic trio of overachievers. The vividly descriptive quotations that follow all come from NASA's history of the Skylab program (Compton & Benson, 1983).

> The crew had run behind schedule for the first 10 days. After the EVA on 6 August, Bean asked Mission Control how far they had fallen behind. Houston's responses bolstered the astronauts' resolve to catch up; as Lausma recalled, "we decided that we weren't going home without doing 100% . . . and more if possible." The turnaround during the next two

> weeks was striking. Whereas Houston had previously given the crew more than it could handle, flight controllers were soon hard pressed to find enough work. (p. 310)

In retrospect, it seems obvious that the extraordinary performance of this crew was a type of compensatory behavior that could only be expected in a context similar to theirs. Nevertheless, this was the performance and "style" to which the flight controllers became accustomed, and which were to shape their attitudes toward the next crew. Mission Control came to expect 12-hour workdays as a regular routine.

When the third Skylab crew was launched, everyone was hypersensitive about the issue of motion sickness following the Skylab II experience. It had been a problem ever since the larger Apollo modules replaced the smaller Mercury and Gemini modules, but never before had an entire crew become ill. Nausea was once again to be a problem, but this time only indirectly, as it served to signal the powerful cohesiveness of the flight crew and its alienation from the flight controllers.

> After [rendezvous and docking with Skylab] the crew was out of touch with Houston for 41 minutes between Bermuda and Carnarvan, Australia, so they started straightening up the command module, stowing the gear used during rendezvous and docking. First, however, Carr and scientist-pilot Edward G. Gibson took their antinausea pills. Pilot William R. Pogue had already attended to that, but too late. A few minutes before ground contact was established, he asked Gibson to hand him a vomit bag. Gibson complied, and as he and Carr went ahead with their chores, Pogue said, "I think I'm going to go slow for the next few minutes." It was not enough; weightlessness had done its work, and Pogue vomited—not very much, but he was quite nauseated. Houston came back on the communications circuit just before 6 p.m. and reiterated the physicians' warning about entering the workshop. [They were concerned that the extra motion might induce nausea before the crew had a chance to adapt] . . .
>
> After Skylab went out of radio range, Carr and Gibson debated what they should say about Pogue's illness during the evening status report, due in just over an hour. Carr was inclined to keep mum for the time being. To account for the food Pogue had not eaten, Carr would say that Pogue was not hungry. As they prepared their second meal, Carr and Gibson mulled over the situation. . . .
>
> Had they remembered that an onboard tape recorder was running all this time, Carr and Gibson would have reported Pogue's vomiting. But they did not remember; and, thinking that only the three of them would

> ever know what had happened, they decided to minimize the potential repercussions of the pilot's illness. Pogue had vomited very little; it was not a gut-wrenching attack. Surely he would recover before they moved into the workshop the next morning. . . . So during the medical conference, Pogue's nausea was mentioned but not the vomiting. Before retiring, Carr read the evening status report to the ground, reporting that "the pilot had no strawberries for lunch and has not eaten meal C." . . .
>
> Meanwhile, the tapes from the onboard recorder were being routinely transcribed in Houston, revealing the candid discussions Carr and Gibson had held regarding Pogue's illness. Reaction was prompt. A medical conference was called in midafternoon. Toward the end of the day, Alan Shepard, Chief of the Astronaut Office, took the microphone in mission control to give the crew a public and official, if mild, reprimand. "I just wanted to tell you," he said, "that on the matter of your status reports, we think you made a fairly serious error in judgment here in the report of your condition." Carr accepted the rebuke: "Okay, Al. I agree with you. It was a dumb decision." And that was that. At that evening's change-of-shift press briefing, reporters wondered if the incident portended a break in frank and open communication between crew and flight controllers. Flight Director Neil Hutchinson thought not; but if there were any further signs of lack of candor, he said, flight controllers would immediately take steps to set matters right. (Compton & Benson, pp. 314-316)

Total candor did not occur. Furthermore, it was not the controllers who ultimately took steps "to set matters right." Ultimately it was the flight crew that found that the situation was becoming intolerable. About six weeks after lift-off, the following events took place:

> Flight planners, having mastered the complex art of assembling a day's activity for three men without wasting a minute, were justifiably proud of their expertise and the quantity of scientific data it could produce. The astronauts, however, did not share that philosophy; they felt their job was to turn out quality results, not merely some arbitrarily large quantity of data. And they chafed under the inflexible scheduling; every tiny housekeeping chore had its bit of time in the daily routine. All three felt that the flight plans were dragging them around by the nose and that the system was not responsive to their needs.
>
> Around Christmas, Carr, Gibson, and Pogue agreed that they had to have a better understanding with the flight planners as to the way things were done. . . . [Carr] taped a six-minute plea for a frank discussion of the mission's status at the half-way point. "We'd all kind of hoped before the mission" he said, that "everybody had the message, that we did not plan

> to operate at the [previous crew's] pace." Now he was worried about how his crew was measuring up to expectations. He was puzzled by some of the questions being asked; he had begun to wonder, "Are we behind, and if so how far?" Were flight controllers worried because the crew wanted so much free time? Were they upset by the time the crew wanted for exercise? "If you guys think that's unreasonable, I'd like some straight words on that." . . .
>
> Carr later regretted that he had waited so long. "We swallowed a lot of problems for a lot of days because we were reluctant to admit publicly that we were not getting things done right," he recalled. "That's ridiculous, [but] that's human behavior." With that summation both of his crewmates emphatically agreed.
>
> The astronauts were not the only ones who felt they needed a frank exchange of views. Robert Parker recalled that ground personnel too were inhibited by the open communication channel. No one who spoke directly to the crew ever suggested that they were doing less than a great job. "We just very seldom [found] ourselves capable of calling a spade a spade," was the way Parker put it. It got worse when the newspapers began to suggest that the third crew was slower and more error-prone than the second. Everyone in Houston became defensive about the crew, feeling that they were being maligned. (Compton & Benson, pp. 327-328)

Some substantive issues were settled in the discussion between the crew and Mission Control, but its main importance was that the differing expectations of the crew and the ground controllers were finally brought out in the open. Mission Control had not been aware of the crew's intentions to work at a less hectic pace than the second crew nor of their frustration that added tasks deprived them of the time necessary to do the original tasks correctly. Nevertheless, when the flight planners added up the accomplishments of Carr's crew as compared with those of the second crew during days 16 to 30 of their missions, they were surprised to find no essential difference. Though the planners still hinted that they would like a heavier workload, an agreement was reached that provided the crew with some control over the timing of less critical activities, such as housekeeping tasks, as well as some free time to unwind at the end of the day.

The events described above took place in a simple space station with a 3-man military crew during the first half of an 84-day mission. In many ways, however, they were predestined by events that occurred, or failed to occur, before lift-off. The technology that made all of this possible was marvelous. But what was making this system—including the people and the apparatus—work or not work properly were psychological phenomena: attributions, congruency between expectancies and outcomes, actual and perceived locus of control, group cohesiveness, group memberships, allegiances, constituencies, and conflict resolution techniques. It is easy to see that interpersonal and group-behavior skills training, as suggested by Nicholas (1989), might have been very useful in bringing an earlier resolution to problems between the flight controllers and the flight crew.

In addition to intergroup conflict, there are other factors to be considered regarding intra-group behavior. The following comments by Nicholas (1987) illustrate some of these considerations:

> The current selection process does not guarantee effective teamwork. People selected for space flight are the brightest and the best, but there is little in their selection or training that concerns interpersonal and team skills. Matters may turn worse as more of the crew become payload specialists (Jones & Annes, 1983). Whereas test pilots are accustomed to working in life-and-death situations, scientists, engineers, and others are not, and may react in panicky or maladaptive ways which interfere with teamwork and jeopardize crew safety in emergencies. Even under routine conditions, heterogeneity is a source of intra-crew conflict. High motivation will carry a crew through conflict on brief missions, but interpersonal problems can surface on longer missions, and might compromise group performance. (p. 1011)

Current thinking about the length of a tour of duty for space station crews favors a 90-day cycle. There are two schools of thought about how this should be arranged. One plan is to have crew members train together and then go up to replace the entire crew that is reaching the end of its 90-day cycle. This has the advantage of keeping the crews homogeneous but lacks a sense of continuity. Another plan calls for 90-day cycles, with half of the crew changing each 45 days. In such an arrangement, at least half of each crew would have 45 days or more of experience at any time. One disadvantage of this plan comes from breaking up one group and forming a new one that has not "proven itself" during training. There is also the medical disadvantage that the group isolated on the space station will have developed and adapted to a very stable biosphere. The introduction of new people from a very different biosphere is very likely to produce periods of infection (e.g., colds, flu) shortly after a crew change.

Ecological psychology (see Wicker, 1979) is a potentially rich source of research concepts and techniques that seem highly relevant in studying space environments. Yet in all of the space literature I have read I have seen no references made to it. Two particular concepts from this branch of social psychology strike me as being especially germane in space studies: the notion of *behavior settings* and their self-regulating character, and *manning theory*, which bears on the motivational consequences of having too few, too many, or just the right number of persons to accomplish a given task.

Finally, in thinking about issues of group dynamics, one must realize that long-term spaceflight with crews of mixed sexes will introduce serious social concerns that have not been dealt with in the space program. Some scientists, such as Yvonne Clearwater at NASA's Ames Research Center, have begun to raise these issues (see Clearwater, 1985).

Cross-Cultural Issues

The space station Freedom is often referred to as an international space station. While the major impetus for its design and development comes from the United States, the completed structure will be a collaborative effort. The United States will design and build the superstructure, habitat, and one of the three laboratory modules. One of the laboratory modules will be designed and built by Japan and another by the European Space Agency (ESA). Finally, the large robotic manipulator arms, similar to the manipulator on the shuttles, will be provided by Canada. When the station is fully occupied, it will probably have an 8-person crew. Eventually, one might anticipate a multinational crew of the following sort: a U.S. commander and two U.S. crewpersons, a Canadian, two Japanese, and a Frenchman and a German representing ESA. If we have already had intergroup problems involving a crew only comprised of Americans, imagine the difficulties that might arise when different constituencies and cultures make up a crew. Suddenly there will be cultural differences in foods and times for eating to be considered. Similarly, proxemic differences in the size of crew members' personal space are to be expected. For instance, Arabs tend to stand closer together when they speak to someone than do Americans. On earth, when an American converses with an Arab, the American tends to step backwards to adjust the spacing to a "proper" distance, whereupon the Arab takes a step forward to accomplish the same goal. Of course, this is usually accomplished without either party being aware of what is happening, and the Arab is left wondering why Americans never stand still when they converse (Hall, 1969; see Altman & Chemers, 1980, for an excellent discussion of culture and environment). Such differences also exist in cultures less diverse than those of the United States and Arab nations. There are proxemic differences of Germans and Americans in larger settings, such as rooms, of which Hall wrote: "None of these American spatial definitions is valid in northern Germany. In every instance where the American would consider himself *outside* he has already entered the German's territory and by definition would become involved with him" (p. 133). Clearly, both the physical design and layout of the space station—as well as the rules of proper conduct within it—must take into account cultural differences among the likely inhabitants.

One fascinating cross-cultural issue has already arisen on one of the shuttle missions. On a 1985 flight of Discovery, Sultan Salman Abdelazize Al-Saud of Saudi Arabia served as a mission specialist to witness the launch of an ARABSAT 1-B communications satellite. Because the sultan was a Moslem, it was necessary for him to face toward Mecca and pray at certain times of the day. In the U.S., this is a simple task—when the proper time of day arrives, one faces the east and prays. However, in low earth orbit, the sun rises and sets every 45 minutes, making a solar day only 90 minutes long. The solution to the problem of when to pray was to use Houston time as the reference time for the space crew. Thus, when it was prayer time in Houston, it was prayer time onboard the shuttle. The next problem involved the direction to face while praying: Which way was east? Because there is no east in space, the problem was solved when religious authorities decided that facing toward the earth was the appropriate behavior. These were sensible and relatively easy decisions to make. What is most important is that it was recognized that such decisions had to be made, and the very important religious needs of a crew member were anticipated and dealt with before the flight took place.

Groups under threat tend to develop strongly cohesive bonds. Space crews are under constant *physical* threat from their hazardous environment, as well as being under *social* threat due to their high visibility before the whole world. We have already seen the consequences of this in the previously described Skylab events. To suggest some of the complexity of personal, political, managerial, contractual, and financial dynamics that are likely to take place on the future space station, let me present the following imaginary scenario.

Suppose there is a crew made up of a U.S. commander and two other U.S. crew persons, a Canadian, a German, a Frenchman, and two Japanese. They have been in space for 40 days. They have faced a couple of minor emergencies that were capably handled because the crew was highly competent and performed emergency responses in a very cooperative manner. On one occasion, the German and French crew members (two of the four crew members qualified and outfitted for EVA work) donned space suits and spent 6 difficult hours outside the pressure vessels removing an experiment package from the external palette of the Japanese-built laboratory after it had become jammed and could not be returned through the air lock by remote control.

An experiment, sponsored by a German firm, has been underway in the ESA-built laboratory. This experiment requires 15 kilowatts of electrical power. The complex and detailed contractual time line developed by NASA calls for that amount of power to be devoted to the German experiment, after which it is scheduled to be shifted to a Japanese experiment and then to an experiment paid for by a U.S. firm. Fifteen minutes after the power was transferred to the German experiment and the project was begun, the German crew member realized an error had been made, and the 2-hour project had to be started over. Now, 45 minutes after starting over, everything is moving along smoothly. In exactly one hour, the power is scheduled to be diverted to the Japanese firm's experiment. The German scientist approaches the U.S. commander, explains the problem, and requests to be allowed to keep the 15 kilowatts of power for an extra 15 minutes to finish the experiment, explaining that without those 15 minutes of power the experiment will be ruined. The scientist is contrite and embarrassed by the error and hopes the commander will not "broadcast this to the whole world" because "we've all made mistakes up here from time to time."

Suppose you were the commander. What would you do? Who should have the authority to make such a decision—the commander, the flight crew as a group, the flight controller in charge of that shift, the chief flight controller, the financial officer responsible for contracts, the controller in charge of writing the time line, or the head of NASA? There isn't much time left. A decision must be made.

Clearly, it will be best if issues such as this one are anticipated and studied in advance, and the mechanisms for coping with them are worked out by utilizing the best information available from fields within the social sciences. However, one can see how concern about problems like this must seem premature to an agency such as NASA or its contractors, especially as they have not yet worked out many of the design problems for the structure of the space station and other mechanical components, such as an emergency return vehicle.

Looking Ahead

I have described many of the ways in which modern technology in the space program is affecting traditional behavior and producing tantalizing challenges for applied social psychologists. Ultimately, applied social psychologists will have to work more intimately with design and engineering professionals from the time of the very first discussions about a design prototype. For this interaction to be effective, social scientists will have to increase their technological literacy, and engineers will have to reciprocate by becoming more literate in such areas as human factors. Hardware contractors will need to increase the number of social scientists that are a part of their design teams. However, before such phenomena occur extensively, the current crop of applied social scientists will have to provide, through research, data showing what the needs are and suggesting the way to their solution. I will conclude this chapter by proposing some avenues that I think lead in the right direction.

Given the enormous cost of putting a person in orbit, it would be difficult to convince NASA to put social scientists up in the shuttles or in Spacelab to do on-site field studies of the astronauts. In fact, that may not be desirable or necessary from an experimental design point of view. With current videotape technology, most of the crucial areas of interpersonal behavior and the person-machine interface occurring on a flight can be studied from earth. Once all of the tracking and data relay satellites are in place, orbiters will be in constant contact with the ground, and social scientists at a control center will be able to have real-time control over such things as camera position, lighting, focus, and zoom. Much better quality images can then be obtained, and studies can be designed that will use various types of time-sampling techniques to select data from the total set of behaviors that occurred. This research would represent a very small percentage increase in the total cost of the space program, and also would interfere very little with other work taking place.

The major problem with this research would be the management of privacy for the space crews. It might be necessary for every flight to have some time that was free of surveillance. However, this is no more of a hurdle than those that have to be dealt with when research is conducted in field settings. If the social scientists make the flight crews sufficiently aware of the importance of this research, the crews ought to be eager to cooperate in the acquisition of the data.

Another mechanism for performing useful research before a space station exists, or on earth without using expensive space equipment, is through simulation in spacecraft mock-ups. Of course, the zero-g variable is not present under such conditions, nor are such factors as danger and true isolation. Still, most of the variables having the strongest effects on social interactions could be included, and much valuable research could be conducted. NASA has been a leader in the development of simulation techniques using mock-ups, which are used extensively in planning for each flight into space. However, this technique has not yet caught on with academic social scientists (who train the new generation of scientists), partly because such scientists are remote from the NASA facilities and because mock-ups are expensive to build. A rather significant number of mock-ups exists around the country at various NASA facilities and at contractor sites; often they sit idle for long periods or, after being used to support engineering and design, they become museum pieces under the auspices of public relations departments. Many of these facilities could be utilized when there is sufficient interest on the part of the applied social science community to get involved and propose research projects.

Simulations often turn out to be surprisingly true to life, and they have been used effectively by airlines and NASA, and in psychotherapy and a variety of social psychology research (e.g., Gamson, 1978; Guetzkow, Kotler, & Schultz, 1972; Raser, 1969). In fact, most laboratory studies in small-group dynamics, leadership, communications, management, and conformity incorporate some type of simulation in order to increase the external validity of the research.

Applied simulation research can be usefully employed to plan ahead for contingencies that might be faced in spaceflight. An example might be the hypothetical scenario presented above in which a scientist made an error and needed electrical power for an additional 15 minutes. By setting up simulations, researchers could study different solutions, management approaches, communication patterns, and other variables involved in coping with conflicts and emergencies. By having astronaut candidates from the different constituencies perform in space station simulators, very high-fidelity simulations could be achieved. These could help in planning for specific situations that might arise, and, as often happens in good applied research, they would be likely to generate hypotheses of a more basic nature that could then be further explored and tested. The results would probably have a high payoff in generality and produce spin-off information useful in other severe environmental settings.

The information of artists has already come into play through the scenarios they pen in science fiction stories. In his book, *Welcome to Moonbase*, Ben Bova (1987), president of the National Space Society, provided a captivating look at life on a futuristic moon base in the early twenty-first century. This story is rich in technical accuracy, plausibility, and well-thought-out social psychology. I predict that the day is not far off when people will look back in amazement at Bova's prescience. Once again, it is time for social scientists to climb on the bandwagon to the future and catch up with the artists.

Great excitement awaits social scientists who get involved in space research and applications. They will be awed frequently by the enormity of the unforeseen challenges associated with living away from the planet upon which we evolved. In light of the fact that some quite advanced engineering is well underway toward the construction of our next space station, there is some urgency for a dramatic increase in the space efforts of social scientists.

References

Altman, I., & Chemers, M. M. (1980). *Culture and environment*. Monterey, CA: Brooks/Cole.

Banks, P. M., & Ride, S. K. (1989). Soviets in space. *Scientific American*, *260*, 32-40.

Bluth, B. J., & Helppie, M. (1986). *Soviet space stations as analogs* (2nd ed.) (NASA Grant NAGW-659). Washington, DC: NASA.

Bova, B. (1987). *Welcome to moonbase*. New York: Ballantine.

Clarke, A. C. (1968). *2001, a space odyssey*. New York: New American Library.

Clearwater, Y. (1985, July). A human place in outer space. *Psychology Today*, pp. 36-43.

Compton, W. D., & Benson, C. D. (1983). *Living and working in space: A history of Skylab*. Washington, DC: NASA.

English, H. B., & English, A. C. (1958). *A comprehensive dictionary of psychological and psychoanalytical terms*. New York: David McKay.

Foushee, H. C. (1982). The role of communications, sociopsychological, and personality factors in the maintenance of crew coordination. *Aviation, Space, and Environmental Medicine, 53*, 1062-1066.

Foushee, H. C. (1984). Dyads and triads at 35,000 feet: Factors affecting group process and aircrew performance. *American Psychologist, 39*, 885-893.

Gamson, W. A. (1978). *SIMSOC: Simulated society. Participant's manual* (3rd ed.). New York: Free Press.

Grossman, S. P. (1967). *A textbook of physiological psychology*. New York: John Wiley.

Guetzkow, H., Kotler, P., & Schultz, R. L. (1972). *Simulation in social and administrative science*. Englewood Cliffs, NJ: Prentice-Hall.

Hall, E. T. (1969). *The hidden dimension*. Garden City, NJ: Doubleday.

Jones, D. R., & Annes, C. A. (1983). The evolution and present status of mental health standards for selection of USAF candidates for space missions. *Aviation, Space, and Environmental Medicine, 54*, 730-734.

Kanas, N. (1985). Psychosocial factors affecting simulated and actual space missions. *Aviation, Space, and Environmental Medicine, 56*, 806-811.

Kanas, N. (1988). Psychosocial training for physicians on board the space station. *Aviation, Space, and Environmental Medicine, 59*, 456-457.

Kinsey, J. L. (1959). Psychologic aspects of the "Nautilus" transpolar cruise. *Armed Forces Medical Journal, 10*, 451-462.

Loftus, J. P., Jr. (1986). Space: Exploration-exploitation and the role of man. *Aviation, Space, and Environmental Medicine, 57* (Suppl. 10), A69-77.

McClellan, J. M. (1985, December). Return to earth. *Flying*, pp. 34-38.

McGinnies, E. (1949). Emotionality and perceptual defense. *Psychological Review, 56*, 244-249.

Miller, J. W., VanDerwalker, J. G., & Waller, R. A. (Eds.). (1971). *Tektite 2: Scientists-in-the-sea*. Washington, DC: U.S. Department of the Interior.

Mullin, C. S., Jr., & Connery, H. J. N. (1959). Psychological study at an Antarctic IGY station. *Armed Forces Medical Journal, 10*, 451-462.

Nicholas, J. M. (1987). Small groups in orbit: Group interaction and crew performance on space station. *Aviation, Space, and Environmental Medicine, 58*, 1009-1013.

Nicholas, J. M. (1989). Interpersonal and group-behavior skills training for crews on space station. *Aviation, Space, and Environmental Medicine, 60*, 603-608.

Raser, J. R. (1969). *Simulation and society*. Boston: Allyn & Bacon.

Santy, P. A. (1987). Psychiatric components of a health maintenance facility (HMF) on space station. *Aviation, Space, and Environmental Medicine, 58*, 1219-1224.

Shapland, D., & Rycroft, M. (1984). *Spacelab: Research in earth orbit*. London: Cambridge University Press.

Thornton, W. E., Hoffler, G. W., & Rummel, J. A. (1977). Anthropometric changes and fluid shifts. In R. S. Johnston & L. F. Dietlein (Eds.), Biomedical results from Skylab (NASA SP-377). Washington, DC: National Aeronautics and Space Administration.

Vaughan, E. H. (1947). Adjustment problems in a concentration camp. *Sociology and Social Research, 32*, 513-518.

Wichman, H. (1987a). Using speech technology on the space station. *Speech Technology, 4*(2), 20-23.

Wichman, H. (1987b). *Voice recognition and synthesis systems: Their potential for use in the space station* (Tech. Rep. SSS 87-0097). Downey, CA: Rockwell International.

Wichman, H., & Lynch, W. (1988). *A voice integrated controller alerting system.* Unpublished manuscript, report of completed demonstration project. Pasadena, CA: Jet Propulsion Laboratory.

Wichman, H., & Polson, K. (1988). Emotional responses to machine errors following manual vs. voice control. In R. Hopkins (Ed.), *Official proceedings of Military Speech Tech '88: Voice applications for military and government agencies* (pp. 132-127). New York: Media Dimensions.

Wicker, A. W. (1979). *An introduction to ecological psychology.* Monterey, CA: Brooks/Cole.

Author Index

Subject Index

About the Authors

GEZA BOTTLIK earned his bachelor's degree from Manhattan College and holds two graduate degrees in mechanical engineering from the University of Southern California. Since 1960, he has worked for two large electronics manufacturing firms in a wide variety of engineering, research, and factory management positions, most recently including Manager of Factory Automation and Manager of Industrial Engineering.

BARBARA L. CHALFONTE is a behavioral science research intern at Bell Communications Research (Bellcore), where she conducts research on the use of information technology for interpersonal communication. She has a master's degree in cognitive psychology from McMaster University and previously headed a speech research laboratory at the University of Maryland.

DONALD D. DAVIS is Associate Professor of Psychology and director of the doctoral program in industrial/organizational psychology at Old Dominion University. He received his Ph.D. in psychology from Michigan State University, where he also served on the psychology faculty and as an Assistant Director of the Center for Evaluation and Assessment. He has consulted with more than 75 public-sector and private-sector organizations throughout the United States. His chief interests are the management of technological innovation and organization change, and he recently published *Management of Technological Innovation: Organizational Strategies for Implementing Advanced Manufacturing Technologies.*

ROBERT S. FISH earned his Ph.D. from Stanford University and is a cognitive psychologist in the Interpersonal Communications Research Group at Bell Communications Research (Bellcore). His current research is related primarily to problems and systems associated with fostering informal communications between distant coworkers. Some of his recent publications have concerned the Quilt system for collaborative writing and the Cruiser system for multimedia social browsing. He has also worked on evaluating the perceptual components of telecommunications services, particularly the subjective visual quality of transmitted video.

H. CLAYTON FOUSHEE received a B.A. degree in psychology from Duke University in 1975 and a Ph.D. in social psychology from the University of Texas in 1979. He has been awarded fellowships from the National Research Council and the National Institute of Mental Health. Between 1981 and 1989, he conducted human factors research on issues pertaining to crew performance in both aviation and space environments at NASA-Ames Research Center, and initiated a crew factors research program in the Aerospace Human Factors Research Division. Currently, he is the Chief Scientific and Technical Advisor for Human Factors with the Federal Aviation Administration.

BARBARA A. GUTEK is a professor in the Department of Management and Policy at the University of Arizona. Her Ph.D. is from the University of Michigan, with a specialization in organizational psychology. She has served on the psychology faculties of the University of California at Los Angeles and Claremont Graduate School. Her research interests are the effects of computers in the workplace, women and work, sexual harassment at work, and job satisfaction. She is currently working on a multiyear study, supported by the National Science Foundation, regarding relationships among the organization of work in groups, their computer configuration, and work group effectiveness.

BARBARA G. KANKI earned her B.S., M.A., and Ph.D. degrees from the University of Chicago, with a doctoral specialty in cognition and communication. Since 1985, she has worked in the Aerospace Human Factors Division at NASA-Ames Research Center in California. There she has focused on communication processes affecting aircrew performance, training, and flight safety, including research on social/organizational and personality variables in a large range of actual, simulated, and analogous aerospace environments.

ROBERT E. KRAUT is Distinguished Member of Technical Staff at Bell Communications Research (Bellcore), the research arm of the regional telephone companies, where he manages the Interpersonal Communications Research Group. He earned his Ph.D. in social psychology from Yale University, and has held positions at Bell Laboratories, Princeton University, Cornell University, and the University of Pennsylvania. Recently his research has concentrated on social aspects of current and future information technologies, including office automation and employment quality, home-based employment, the communication needs of collaborating scientists, and the design of information technology for small-group intellectual work. He has served on the National Research Council's Committee on Women's Employment and Related Social Issues and on its panel on Technology and Women's Employment. He is currently serving on the Office of Technology Assessment's advisory panel on Information Networks in Research.

ANN MAJCHRZAK is Associate Professor in the Human Factors Department at the University of Southern California (USC). She received her B.A. degree from Pitzer College, and her M.A. and Ph.D. degrees in social psychology from the University of California at Los Angeles. Before joining the faculty at USC, she was an organizational and management consultant and contract researcher in Washington, DC, and then a faculty member in the Krannert Graduate School of Management at Purdue University. Majchrzak has been actively involved in research and consulting on the design and assimilation of advanced manufacturing technology for the past seven years, with support from the National Science Foundation, the Congressional Office of Technology Assessment, several manufacturing companies, and the RAND Corporation. Her books include *The Human Side of Factory Automation* and *Human Aspects of CAD.*

PAUL D. NEWMAN is a Project Director/Analyst at Westat, Inc., in Rockville, Maryland. During his eight years at Westat, he has directed or managed a series of projects on the implementation of new technologies in the workplace, employee survey and communication programs, management support of workplace safety, development and support of entrepreneurship and innovation, sales training and incentives, and the status of research facilities in academic and private research institutions throughout the United States. He holds a master's degree in organizational psychology from Virginia Polytechnic Institute and State University.

VERONICA F. NIEVA is a social/organizational psychologist who has conducted research concerning human resource and organizational change issues in private sector manufacturing and service organizations, domestic federal agencies, and military services. Her work on the implementation of programmable manufacturing technologies stems from a broader interest in studying key changes in the workplace. She obtained her doctorate in organizational psychology from the University of Michigan, and is now Vice President and Director of Organizational and Management Research at Westat, Inc., a social science research firm in the Washington, DC, metropolitan area. She has published two books, *Women and Work* (with B. Gutek) and *Managing Creation: The Challenge of Building a New Organization* (with D. Perkins and E. E. Lawler).

STUART OSKAMP is Professor of Psychology at Claremont Graduate School. He received his Ph.D. from Stanford University and has had visiting appointments at the University of Michigan, University of Bristol, London School of Economics and Political Science, and University of New South Wales. His main research interests are in the areas of attitudes and attitude change, behavioral aspects of energy and resource conservation, and social issues and public policy. His books include *Attitudes and Opinions* and *Applied Social Psychology*. He is a past president of the APA Division of Population and Environmental Psychology and is currently editor of the *Journal of Social Issues*.

ROBERT W. ROOT received his Ph.D. in social psychology from the University of California, San Diego, worked on human factors problems for the Essex Corporation, and is currently in the Multimedia Communications Research Division at Bell Communications Research (Bellcore). He has worked on the design and evaluation of new services concepts, including service and interface definition for the MICE (Modular Integrated Communications Environment) services prototype test bed, and the creation, development, and evaluation of graphics interfaces for end-user network control. He is currently working on the Cruiser project—an advanced audio-video communications system designed to provide desktop support for distributed multiparty work groups.

SHIRLYNN SPACAPAN is Associate Professor of Psychology at Harvey Mudd College and at the Claremont Graduate School. She received her Ph.D. from the University of Oregon in 1982, where she taught for two years before moving to Claremont. She is currently studying topics at the interface of environmental psychology and organizational behavior, such as perceived control in the workplace and workplace design. Recently she held a Haynes Foundation Faculty Fellowship for her work on the social psychology of aging.

HARVEY WICHMAN received his B.A. and M.A. degrees from California State University, Long Beach, and his Ph.D. in experimental psychology from Claremont Graduate School. He has held faculty positions at Delta College; California State University, San Bernardino; and Claremont McKenna College, where he currently chairs the Department of Psychology and directs the Aerospace Psychology Laboratory. A graduate of NASA's Bio-Space Technology Training Program, he is an active commercial pilot and flight instructor. He recently spent a year as a Sloan Foundation Fellow with Rockwell International as part of the engineering team developing the preliminary design for the international space station, *Freedom*. While working at the Jet Propulsion Laboratory in the summer of 1987, he developed the voice interactive system described in this volume.

SUSAN J. WINTER received her master's degree in organizational psychology at Claremont Graduate School in 1989 and is a Ph.D. candidate at the University of Arizona in the School of Business Administration. She is currently pursuing research on computer-using employees' affective reactions to work. She is also interested in cognitions about computer use, end-user and desktop-user computer expertise, and norms of time in the workplace.